ENERGY AND ECONOMIC GROWTH IN THE UNITED STATES

The Institute for Energy Analysis and The MIT Press
Perspectives in Energy Series
Alvin M. Weinberg, general editor

*Economic and Environmental Impacts of a Nuclear Moratorium,
1985-2010, Second Edition,* by the Institute for Energy Analysis,
1979

Energy and Economic Growth in the United States by Edward
L. Allen, 1979

ENERGY AND ECONOMIC GROWTH IN THE UNITED STATES

Edward L. Allen
Institute for Energy Analysis
Oak Ridge Associated Universities

The MIT Press
Cambridge, Massachusetts, and London, England

Oak Ridge Associated Universities is a private, not-for-profit association of 46 colleges and universities. Established in 1946, it was one of the first university-based, science-related, corporate management groups. It conducts programs of research, education, information, and training for a variety of private and governmental organizations. ORAU is noted for its cooperative programs and for its contributions to the development of science and human resources in the South.

The Institute for Energy Analysis was established in 1974 as a division of Oak Ridge Associated Universities to examine broad questions of energy policy. More specifically, it assesses energy policy and energy research and development options and analyzes alternative energy supply and demand projections from technical, economic, and social perspectives. The Institute focuses primarily on national energy issues, but it is also concerned with international energy questions and their implications for domestic energy problems.

This volume is based on work performed under contract between the National Research Council and Oak Ridge Associated Universities. Oak Ridge Associated Universities operates under contract number EY-76-C-05-0033 with the U.S. Department of Energy.

First MIT Press edition, 1979

Printed and bound in the United States of America

Library of Congress Cataloging in Publication Data

Allen, Edward Lawrence, 1913-
    Energy and economic growth in the United States.

    Includes bibliographical references and index.
    1. Power resources—United States. 2. Energy policy
—United States. 3. United States—Economic conditions
—1971- I. Oak Ridge Associated Universities.
Institute for Energy Analysis. II. Title.
HD9502.U52A42 333.7 79-15549
ISBN 0-262-01062-3

ACKNOWLEDGMENTS

2131125

IEA wishes to thank the many experts consulted in the preparation of this study. These include Jack Alterman, Peter L. Auer, Hans Bethe, Calvin L. Burwell, Monte Canfield, Jr., Roger S. Carlsmith, Guy Caruso, Thomas B. Cochran, Floyd L. Culler, George Daly, Edward F. Denison, William Fulkerson, Howard Fullerton, Campbell Gibson, Walter R. Hibbard, Jr., L. John Hoover, Kenneth A. Hub, Philip L. Johnson, Michael Kennedy, George B. Kistiakowsky, Ronald Kutscher, Hans Landsberg, Beverley J. McEaddy, Dennis L. Meadows, Robert Miki, Celia Evans Miller, Philip F. Palmedo, Robert G. Sachs, Sam H. Schurr, John H. Vanston, Jr., David R. Weinberg, Larry J. Williams, and Louis Wofsy. The author thanks John Gehman for the statistical work and Rayola Dougher for typing the manuscript. Of course, none of them are in any way responsible for our findings—this responsibility is borne solely by IEA.

INTRODUCTORY NOTE

Energy and Economic Growth in the United States is the second volume in the Institute for Energy Analysis–MIT Press series entitled *Perspectives in Energy.* It is a companion to the first volume, *Economic and Environmental Implications of a U.S. Nuclear Moratorium, 1985-2010,* which was published in 1976. The moratorium study was based on estimates of future energy demand; the present study explains in detail how the Institute for Energy Analysis arrived at these estimates.

The relation between energy and economic well-being is central to energy analysis. This relation was first examined in detail in the book *Energy in the American Economy, 1850-1975* by Schurr, Netschert, Eliasberg, Lerner, and Landsberg that appeared almost 20 years ago. Remarkably, the estimated aggregate energy demand for 1975 given in this book agreed almost perfectly with the actual energy used in that year. Yet the methods used in making these estimates were rather qualitative and bore little resemblance to the "hard" econometric analysis that now pervades the literature of future energy predictions.

The methods used in this volume are much closer to those used in the early study of Schurr *et al.* than they are to, say, the very elaborate econometric modeling of the Project Independence study. One cannot claim that the use of these methods confers on these estimates a greater likelihood of accuracy. Nevertheless, the methods used in this volume have the advantage of transparency: one can see at each stage precisely what assumptions are being made and what inferences are being drawn from these assumptions. In this respect, I find these semiquantitative methods of prognostication more appealing than the more heavily econometric schemes that are in such vogue now. Too often the entire econometric analysis hangs on a single number—typically, the elasticity of demand for liquid fuels; and the results are correspondingly fragile, since demand elasticities are notoriously unreliable.

The publication of *Energy and Economic Growth in the United States* in 1979, some 3 years after much of the underlying work was completed, has a disadvantage: one can begin to compare the predicted and actual turn of events. As for total energy demand, the low estimates seem to be tracking the actual trend. On the other hand, the original estimates for nuclear power have been much too high; this striking change is reflected in this volume but not in its earlier companion moratorium study.

Dr. Allen has done a substantial service for the community of energy analysts in creating order from a mass of seemingly unrelated and disordered data. He has acknowledged most gracefully his debt to his colleague, Dr. Charles Whittle, Assistant Director of the Institute for Energy Analysis, who formulated the semiquantitative scheme for projecting energy demand that has been applied so well by Dr. Allen. I wish to extend my thanks to Drs. Allen and Whittle, as well as to the other members of the Institute for Energy Analysis, who in this volume have made a useful contribution to the literature of energy analysis.

Alvin M. Weinberg
Director
Institute for Energy Analysis
Oak Ridge Associated Universities
Oak Ridge, Tennessee 37830
October 1978

CONTENTS

TABLES

FIGURES

SUMMARY

New Findings Since the 1976 Report

Since the Institute for Energy Analysis (IEA) released its initial report on
U.S. Energy and Economic Growth, 1975-2010, in 1976, a number of supplementary reports have been completed and their conclusions are embodied in this book. These changes and additions are as follows:

1. We have now identified that perhaps the single greatest uncertainty in our earlier and current estimates is the size of illegal immigration. If the net influx of illegal immigrants is approximately 10^6 a year, a number some observers believe corresponds to present experience, the U.S. economy might grow at higher rates than those estimated here. This potential uncertainty has been analyzed and is reported in Appendix A.

2. At the time of our earlier report, coal seemed assured of a larger and rapidly growing share of the public utility and industrial fuel markets. This outcome is now much less certain, even though the long-term competitive position of coal appears to have improved significantly. Costly federally mandated pollution controls and a host of federal laws dealing with mining regulations, combined with sharply higher labor costs, have dulled coal's competitive edge.

3. Detailed regional energy and economic estimates for the year 2000 are included for the 101-quad case. There is great regional variation in energy use per capita; the largest differences are due to the concentration of energy-

intensive industries in a few locations. Other factors are variations in climate and population density. Four regions, the West South Central, East North Central, Middle Atlantic, and South Atlantic, will account for 67 percent of the anticipated national total in 2000 (see Chapter 5).

4. Having subsequently analyzed the impact on energy demand of an aging population and of rising employment in the service industries in more detail, we do not now concur in the commonly held views that these developments are likely to lower the aggregate ratio of energy use to the gross national product (GNP) over the next few decades. We now conclude that the energy impact of an aging population will be toward higher per capita consumption of energy, largely because these adults are in the active work force which constitutes the highest per capita energy-consuming group. They will account for 60.4 percent of the total population in 2000 compared to 53.6 percent in 1975.

5. A detailed examination of the probable growth rates and projected energy efficiency improvements of several major energy-consuming manufacturing industries has led to a much lower estimate of industrial energy demand, some 44.4 quads of total industrial demand in 2000 compared to 50.3 quads estimated in the 1976 study. The earlier estimates of likely household and commercial energy savings by 2000 are believed to have been overstated by perhaps 20 percent.

The Energy Problem

In the historical sweep of the economic growth and development of the United States, energy has been considered a nearly ubiquitous good—essential, to be sure, but abundant and inexpensive. Yet, as the decade of the 1970s opened, there were some warning signals that the era of abundance was coming to a close. Petroleum production in the United States peaked in 1970. The output of domestic natural gas, which supplied the largest share of energy for the economy, peaked in 1972. The embargo imposed in late 1973 by the Organization of Petroleum Exporting Countries (OPEC) on oil exports to the United States dramatically signaled the end of the era of self-sufficiency. The effectiveness of the embargo was made possible by a basic change in America's domestic energy industries, from a comfortable surplus of supplies to a growing dependence on imports. Energy prices, led by oil prices, escalated rapidly.

At the same time, official and private forecasts of future energy demands, based largely on recent historical growth rates of 3.7 percent a year, moved upward relentlessly. The consequence was the predicted emergence of a gap between world oil production and requirements by the early 1980s, which was expected to grow to alarming proportions by the end of the 1980s. If these estimates were correct, economists in the U.S. Department of Energy and the Central Intelligence Agency believed that U.S. economic growth would be severely affected because of a shortage of essential energy.

One of the major conclusions of the IEA studies is that, although there is a serious energy problem, it can be eased by conservation and the stimulation of new sources of supply. Therefore, even though energy prices seem certain to rise, both absolutely and relatively, we have not been able to identify an inevitable supply/demand "crunch" which will produce economic disruption and record high unemployment in this century.

Specifically, this study estimates economic growth (GNP) and energy demand for the United States to the year 2000. We find that the GNP and total energy demand are likely to grow more slowly than has been forecast in most analyses of energy policy sponsored by the U.S. government. Instead of basing our estimates on economic growth rates that are tied to highly optimistic full employment goals, our aim has been to construct what we believe are the most likely economic scenarios and the related future energy needs. Thus, the estimates that emerge from our analysis are in no sense "normative"; we have avoided suggesting what *ought* to be the U.S. energy future. Rather, our estimates flow from an analysis of what we believe is *likely* to happen in a surprise-free world. As has been generally noted, differences in economic growth assumptions exert large effects on calculations of energy requirements.

Many factors point to a lower economic growth rate in the next 20 to 25 years than this country has enjoyed since the onset of World War II, as is explained in Chapter 1. These factors include the sharp drop in the fertility rate during the last decade, which will cut the growth of the labor force by 50 percent by the end of the 1980s in the absence of massive immigration, and a drop in productivity and productive investment that will make productivity gains that occurred in the past more difficult to achieve. Lower energy demands are likely to occur as a result of reduced economic growth, the gradual introduction of energy-saving technologies, and the expected higher energy prices. Although accurate prediction of the future is clearly impossible, many of the underlying factors that will bear heavily on economic growth and energy demands in the decades immediately ahead can now be specified.

Throughout this study we have endeavored to explain precisely the methods used and the reasons for arriving at given estimates. Each reader can then decide for himself the extent to which he would accept or modify the analysis. Long-term energy and economic projection is not a pure science but rather an art based on economic and technological assessments and reason.

Methodology

A summary of our methodology is given in the next several paragraphs. The U.S. energy demands are divided into four broad sectors—households, commercial space, the transportation of persons and goods, and industry. We determined the future growth of energy demands in each sector by combining demographic-economic assumptions with attainable technical efficiencies in specific energy-consuming devices and calculated the rates for the introduction of these newer technologies. The specific energy demands obtained from an analysis of each sector were then summed to obtain the total energy demand.

We arrive at our estimates of energy demand in four specific steps. First, we estimate the GNP (in Chapter 1) by using a simple formulation: GNP equals labor force multiplied by labor productivity. The employed labor force and the hours worked are estimated from projections of population and labor participation rates. The adult population is already determined for much of the period (up to the early 1990s), and the labor participation rate (the number of persons 16 years and older who are either employed or actively seeking employment compared to the total number of persons 16 years and older) is assumed to continue its long-term growth. Labor productivity (the output per worker) is estimated by extrapolation of historic trends. In general, we have tried to bias our results toward the high side. For example, we have used optimistic assumptions about future labor productivity. In Chapter 2, we have been conservative in our judgments regarding future energy conservation.

From projections of the GNP and population, we derive estimates of the size of intermediate factors leading to the energy demand calculations shown in Chapter 3. That is, we calculate the number and type of households expected in the year 2000, the square feet of commercial space needed to support these households, and the number of automobiles expected to be on the road. Since automobile use is a consequence of life-style decisions, which are uncertain, we project two possible trends. One of these, used in the low

(101-quad) scenario, assumes that automobile usage has reached a point of saturation in relation to the population of driving age and that the annual mileage per automobile will remain at 1975's 10,000 miles. The second assumption, used in the high (126-quad) scenario, is that the automobile stock will increase from the present 0.67 car per person over 16 years of age to 0.77 by the year 2000. In addition, the annual mileage per vehicle is allowed to increase from 10,000 to 12,000 miles in 2000.

From the magnitudes of each intermediate factor, we estimate its corresponding end-use energy demand; the total energy demand is then the sum of the energy demands in each end-use category. Two parameters enter into these estimates: the rate of introduction of new technologies (for example, lightweight automobiles) and the degree of energy conservation (the so-called efficiency improvement index). Our projections of population, GNP, and energy demand are shown in Table 1.

We have given independent estimates of energy prices, based generally on extrapolation and judgment instead of explicit prices of energy. Implicit in

Table 1.

Population, GNP, and energy demand estimates for the low (101-quad) and high (126-quad) scenarios

	Totals					
	Population (x 10^6)		GNP (x 10^9 $1972)		Energy (quads)	
Year	Low	High	Low	High	Low	High
1975	213	213	$1192	$1192	71	71
1985	229	231	1730	1730	82	88
2000	246	254	2620	2648	101	126
	Per capita values					
	GNP ($1972)		Energy demand (x 10^6 Btu)			
1975	$ 5,596	$ 5,596	334	334		
1985	7,555	7,490	360	381		
2000	10,650	10,425	411	496		

our energy demand projections are price elasticities, and we find our calculated elasticities to be well within the range of elasticities obtained in other studies.

Assumptions

Given the unknowability of the future, we chose to estimate energy demand according to two scenarios, low and high. The assumptions underlying the two scenarios for each of the elements that determine energy demand are summarized in Tables 2, 3, and 4. Table 2 lists the assumptions made for the key factors that contribute to the growth and composition of the population, the labor force, and GNP. (The specific assumptions underlying Table 2 are discussed in detail in Chapter 1.) Table 3 lists the assumptions made for the key factors determining growth in the intermediate factors for households, commercial (service) space, and automobile inventory. Table 4 lists the assumptions made for the changes in the end-use efficiencies for different energy-use categories. Each of the assumptions is represented by values for selected years between 1975 and 2000.

Our analysis begins with a detailed examination of historic trends for the many factors that determine the growth of the GNP and energy demand. One major factor that implies a lower GNP path is the fertility rate (average number of children per female), which has fallen to 1.8 and is likely to continue at approximately that level. Low fertility will contribute to a slower growth of the labor force, in the absence of higher levels of immigration.

Productivity has been growing much more slowly in the 1970s than it did historically. The reasons for the slower growth rate are not fully understood, but the snowballing costs of pollution abatement and safety requirements, which do not contribute output to the GNP in the conventional sense, have been identified as contributing factors. In spite of other pessimistic factors affecting productivity, we have projected an optimistic recovery of productivity rates between now and the year 2000 from a current 1.8 percent annually to 2.6 percent in the years 1985 through 2000.

In the lower estimate of total energy demand for the year 2000, we have assumed a rapid but far from maximum introduction of energy-saving devices and more efficient technologies. In the higher projection, the pace of conservation is more leisurely. We regard both of these improvements as achievable without a change in life-styles.

Table 2.
Summary of key input assumptions for population, labor force, and GNP growth for the high- and low-energy demand scenarios

Year or period	Fertility rate (children/female)		Labor participation rate (workers per persons 16 and over)	Annual growth rate of full-time employment (%)	
	Low	High	(Low and high)[a]	Low[a]	High[a]
1975	1.8	1.8	0.61		
1975-1985				1.6	1.6
1985	1.7	1.9	0.625		
1985-2000				0.8	0.8
2000	1.7	1.9	0.635		

Period	Annual growth rate of average labor productivity (%)		Annual growth in GNP (%)	
	Low	High	Low	High
1975-1985	2.0	2.0	3.8	3.8
1975-2000	2.6	2.6	2.8	2.8

[a]Immigration in each case is assumed to be the same (400,000 per year) as recent Bureau of the Census population projections. The unemployment rate is assumed to fall to 5 percent.

Table 3.
Intermediate factors for households, services, and automobiles

Year	Households per adult[a] (low and high)	Commercial space per household[b] (square feet) (low and high)	Autos per person over 16 years[c]	
			Low	High
1975	0.53	350	0.67	0.67
1985	0.55	385	0.65	0.71
2000	0.57	437.5	0.65	0.77

[a]Single-person households are assumed to shift to smaller average size housing units.

[b]Commercial composition is assumed to shift from education-type units toward health care and recreation units.

[c]Automobiles are assumed to shift toward lighter-weight vehicles.

Table 4.
Average energy use efficiencies

Year	Households (x 10^6 Btu/unit)		Commercial (x 10^5 Btu/square foot)		Automobiles (miles per gallon)	Truck or bus or rail freight (x 10^3 Btu/ton-mile)	Industrial index (compared to 1975)	
	Low	High	Low	High	(low and high)	(low and high)	Low	High
1975	219	219	3.69	3.69	14	7.1	1.00	1.00
1985					20	6.8	0.85	0.90
2000	214	285	2.82	3.63	27	6.3	0.70	0.80

Findings

Two projections for population, labor force, and GNP are given in Table 5. These projections are based on the analysis in Chapter 1 and the assumptions listed above for future fertility rates, labor participation rates, and labor productivity.

Projections of the number of households, commercial space, and the inventory of automobiles are listed in Table 6. These results are based on the analysis in Chapter 2 and the assumptions for future household formation rates, commercial space and type, and automobile ownership and use listed above.

The key finding is that energy demand over the next 25 years is likely to grow more slowly than in the past and that the ratio of energy use to the GNP will be improved. We also find that the demand for electricity is likely to rise faster than the total demand for energy. Table 7 presents energy demand by sectors. Certain more important findings may be summarized as follows:

1. Long-term average U.S. economic growth after 1985, in terms of real GNP, is likely to be in the range of 2.6 to 3.0 percent annually, even with optimistic assumptions about future growth in labor productivity, unless the flow of illegal aliens into the labor force is very high. This compares to an average annual rate of growth of 3.4 percent for the GNP during the past 35 years.

2. Future long-term growth in U.S. energy demands, even with moderate assumptions about conservation, is likely to be in the range of 101 to 126 quads by the year 2000 if net average energy prices increase at an anticipated annual rate of 2.3 to 4.3 percent and the price increases are gradual and anticipated.

3. The projected growth in the GNP implies that the per capita GNP growth will range from 2.4 to 2.6 percent annually, compared to a growth rate of only 1.8 percent over the past 35 years. The projected annual growth in per capita energy use will range from 1.0 to 1.7 percent compared to 1.4 percent for the past 35 years.

4. Energy-demand scenarios developed here imply a shift to a greater use of electricity, from a current 28 percent of the total to over 46 percent by the year 2000.

5. An analysis of future energy prices and elasticities produces the values shown in Tables 8 and 9.

Table 5.
Population, labor force, and GNP growth

Year or period	Population (x 10^6)		Labor force (x 10^6)	GNP (x 10^9 $1972)		Annual growth in GNP (%)
	Low	High	(low and high)	Low	High	(low and high)
1975	213	213	95	$1192	$1192	
1975-1985						3.8
1985	229	231	111	1730	1730	
1985-2000						2.8
2000	246	254	124	2620	2648	

Table 6.
Households, commercial space, and automobiles

Year	Households (x 10^6) (low and high)	Commercial space (x 10^9 square feet) (low and high)	Automobiles (x 10^6)		Annual automobile mileage (x 10^{12})	
			Low	High	Low	High
1975	72	25.2	104	105	1.05	1.05
1985	87	—	115	125	1.15	1.39
2000	102	44.6	127	152	1.27	1.82

Table 7.
Energy demand by sector
(x 10^{15} Btu)

Year	Total		Households		Commercial		Transportation		Industrial	
	Low	High	Low	High	Low	High	Low	High	Low	High
1975	70.3	70.3	16.2	16.2	9.5	9.5	18.6	18.6	25.9	25.9
1985	82.1	88.0	–	–	–	–	–	–	–	–
2000	101.0	126.0	21.8	29.1	12.6	16.2	22.2	28.1	44.4	52.5

Table 8.

Estimated prices of different
energy modalities
(relative to 1975 in constant dollars)

	1975[a]	1985	2000
Coal	1.0	1.22	1.65
Oil	1.0	1.54	2.40
Gas	1.0	6.42	10.00
Electricity	1.0	1.22	1.65

[a]The 1975 average prices were as
follows: coal, $17.50 per ton, de-
livered to utilities; oil, $10.40 per
barrel, composite cost to refiners;
natural gas, $0.43 per thousand cubic
feet at the wellhead; electricity, 27
mills per kilowatt-hour to the consumer.

Table 9.

Price elasticities[a]
(relative to the reference scenario)

Own price elasticity of demand for energy	Low scenario (101 quads)	High scenario (126 quads)
Coal	−0.23	−0.11
Petroleum	−0.57	−0.19
Natural gas	−0.70	−0.72
Electricity	−0.69	−0.15

[a]The impact of a commodity's own price on demand.

ENERGY AND ECONOMIC GROWTH IN THE UNITED STATES

1

**DEMOGRAPHIC AND
ECONOMIC VARIABLES:
INFLUENCE ON
ECONOMIC GROWTH**

Introduction

In this chapter we discuss the key demographic and economic variables that underlie the IEA analysis of future economic growth in the United States—population, labor force, and productivity (1). Key intermediate factors—households, commercial space, and automobiles—are also examined. Projections of future energy demand, in turn, are based largely on (i) estimates of the growth of these variables and (ii) estimates of the improvement in energy efficiencies over time, which reduce the quantity of energy needed to produce a dollar of output (see Figure 1-1). In Chapter 2 we present estimates of energy efficiency improvements.

Future population estimates, broken down by age groups, are the primary determinants of the sizes, number, and types of households. (Age groups also determine the type of household—single person or family.) The number and types of households are a basic factor determining the residential demand for electricity.

Population growth is largely a function of the fertility rate (FR), defined as the average number of children per female at the completion of her child-bearing years. Net immigration is also a basic factor in population growth estimation, although the flow of illegal entrants has introduced considerable uncertainty into estimates of immigration. The projection of energy demand in

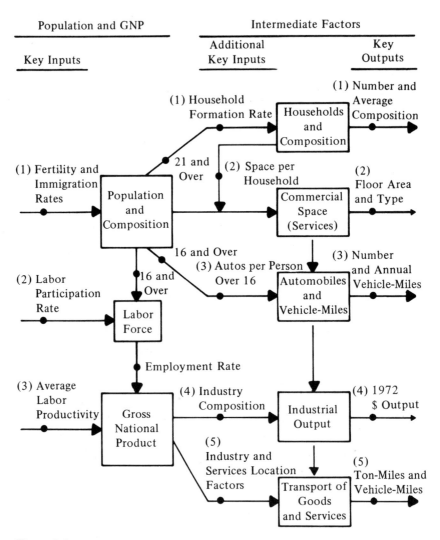

Figure 1-1.
Relationships between population, GNP, labor force,
and intermediate factors

the commercial sector is based on the square feet of service industries' space
needed to support the estimated number of households.

The labor force, by Census Bureau and Department of Labor definition, is
drawn from persons 16 years of age and older. Since only a fraction of this
age group is, in fact, either employed or actively looking for work (others
are housekeepers, students, or other classes of people), a participation rate
(62.1 percent in 1976) is applied to the group over 16 to calculate the size of
the labor force. Participation rates change with time and are projected to increase as a higher proportion of women enter the labor market. The male
participation rate seems to be falling moderately and is expected to continue
to do so unless the change to a 70-year mandatory retirement age alters
recent trends.

The anticipated number of persons 16 years and older also determines the
potential number of licensed drivers and is an essential input into estimating
the future number of personal automobiles in use.

The estimated future growth of the employed labor force multiplied by the
estimated growth in productivity (that is, in output per worker) is the basis
for estimating the gross national product (GNP). Productivity is influenced by
the rate of capital formation, the uses of capital (including the rate of introduction of new technology), and the amount of education per worker. The
GNP growth rate in turn is the prime determinant of the expected increase in
industrial output and in the increased energy needed to transport goods.

The relations between these three basic building blocks—population, labor
force, and the GNP—are shown in Figure 1-1 together with the relations between intermediate factors which they generate and which underlie energy
estimates. Each of these factors will be discussed below. This initial framework of analysis is similar to that used by other economic research groups,
for example, by Department of Labor analysts to forecast occupational
needs [(2), pp. 9-21].

Economic Growth—The Historic Path

Historically, the primary sources of economic growth have been increases in
the inputs of labor and capital and increases in productivity or output per
unit of input. In a series of studies over the past 15 years (3), Edward F.
Denison has estimated the importance of each source of growth for the U.S.
economy. His results are summarized in Table 1-1.

For the period 1929 through 1969, the average growth rate for national income was 3.33 percent; Denison estimated that increases in labor and capital

Table 1-1.
Sources of growth of the total national income
(contributions to the growth rate in percentage points)

	Total actual national income[a]	
	1929-1969	1948-1969
National income	3.33	3.85
I. Total factor input	1.81	2.10
A. Labor	1.31	1.30
Employment	1.08	1.17
Hours	-0.22	-0.21
Average hours	-0.50	-0.37
Efficiency offset[b]	0.19	0.06
Intergroup shift offset[c]	0.09	0.10
Age-sex composition	-0.05	-0.10
Education	0.41	0.41
Unallocated	0.09	0.03
B. Capital	0.50	0.80
Inventories	0.09	0.12
Nonresidential structures and equipment	0.20	0.36
Dwellings	0.19	0.29
International assets	0.02	0.03
C. Energy and raw materials	0.00	0.00

II. Productivity[d] (technical and managerial)	1.52	1.75
A. Advances in knowledge	0.92	1.19
B. Reallocation of resources from farming	0.29	0.30
C. Economies of scale	0.36	0.42
D. Other	-0.05	-0.16

[a]National income differs from the GNP by indirect business taxes and depreciation.

[b]Efficiency offset is the improvement in output per hour worked as a consequence of the decline in the work week. The estimate is arrived at by implication by the method Denison used to measure output.

[c]Intergroup shift offset is a technique for preventing shifts to nonfarm wage and salary workers from farm and nonfarm self-employed groups from decreasing the size of the labor input. Full-time workers in nonfarm wage and salary jobs work many fewer hours per week than the other two groups.

[d]Average labor productivity in the text includes all factors in this table except employment.

contributed 1.81 percent to the economic growth and that increases in productivity contributed 1.52 percent. The increase in the quality and quantity of the labor force has been the most important determinant of the growth rate of the GNP.

In the past, the large supply of raw materials and energy has not been calculated as contributing to growth. Denison (Table 1-1) has made no allowance for raw material and energy inputs in his compilation of sources of growth. In the future, however, shortages of raw materials and energy may result in escalating prices which may inhibit growth at rates comparable to those that prevailed during past years in the United States.

Denison's work is far and away the most detailed study of factors that have contributed to growth in the United States. He provides two tables, one of which is actual growth as in Table 1-1, and the second, which is potential growth. The projections in this IEA study are more closely analogous to Denison's potential growth series. The estimates of potential are more appropriate for long-term growth sources. We have not attempted to modify our projected growth series for future cyclical factors but rather have assumed that the pressure of demand upon available resources is averaged to the year 2000.

Population Estimates

We now begin a series of calculations which lead to the projections of the GNP, a key building block in estimating the quantity of energy needed to run the economy. The steps in the process may be outlined as follows:

1. Estimate the growth of the population to the year 2000 by age groups. This permits us to identify the size of the population which will be of labor force age over time. The size of the population depends on the fertility of American women and on immigration, both legal and illegal.

2. Apply a participation rate to those 16 years and older, in estimating the size of the labor force in the future. The rate for men is higher than that for women, but the fact that female participation is increasing rapidly pushes up the total rate, we think, from 61 percent in 1975 to 63.5 percent in 2000.

3. Subtract the anticipated number of unemployed, since not all persons who are in the labor force are actually employed.

4. Calculate the trend of the average hours worked per week to determine the expected output per employed worker, or each worker's contribution to the GNP. The trend of hours worked, we believe, will continue to decline.

5. Finally, calculate the GNP by multiplying the expected hourly output per worker by the number of hours to be worked. Growth in output per

worker depends upon the growth of productivity. In recent years, since 1970 in particular, productivity per worker has grown much more slowly than in the 1950s and 1960s and may have slowed permanently. The amount of new investment determines the input of modern machinery and equipment in the economy and clearly is an important determinant of productivity growth.

Each of these inputs is discussed below, and Table 1-8 shows the results of our calculations in summary form.

Because there is uncertainty with respect to the growth of the U.S. population to the year 2000, we have used two projections: the Census Bureau Series III rate, an assumed FR of 1.7 children per female, and a rate that is intermediate between the Series III rate and the Series II rate of 2.1, namely, 1.9 children per female (4). Under these alternative assumptions, the estimated populations in the year 2000 would be 246×10^6 and 254×10^6, respectively. Population growth not only determines the size of the future labor force but also is a rough guide to energy demand, since everyone consumes energy directly or indirectly.

The 175-year decline in FR from a level of 7 children per female in the year 1800 to the estimated 1976 level of 1.77 children per female cannot continue for many more decades; if it did, the birth rate would soon approach zero. On the other hand, America's post-World War II "baby boom," which started in the 1940s and continued through 1957, may well turn out to be a unique experience since fertility trends have again resumed their long-term decline (see Table 1-2). We believe the figure of 1.7 to 1.9 children per female reasonably brackets the population prospects to 2000. Changes in life-styles, the general availability of birth control information, rapidly escalating costs of education, and changing moral and ethical standards all seem to be dampening down population growth. The major uncertainty now is the extent to which illegitimate births among teen-agers is likely to decline.

The trend to smaller family size is a general phenomenon in advanced industrial countries around the world, representing in part a change in attitudes among married persons toward childbearing. For example, West Germany's population of about 60×10^6 has already started to decline in absolute numbers (6).

Illegal Aliens and Population Growth
In the long run, aliens entering the United States could be an important component of population growth; this point is quantified in Appendix A. The Census Bureau projections of population include an annual allowance of

Table 1-2.

U.S. total (completed) fertility rates[a]
(children per female) (5)

Year	Fertility rates
1940	2.21
1946	2.83
1950	3.03
1955	3.52
1957	3.76
1961	3.63
1968	2.48
1971	2.28
1974	1.86
1975	1.80
1976	1.77

[a]Completed fertility rate means the
average number of births per female
over a lifetime.

400,000 net immigrants [(4), p. 17]. Actual annual immigration, presumably
all legal, for the years 1972 through 1976 averaged 321,000, excluding the
emergency entry of 130,000 Vietnamese refugees in 1975 (7). The belief of
those at the Census Bureau is that this quantity overstates the number of legal
net entrants, since it does not allow for all emigrants leaving the United States
[(5), p. 17]. Perhaps, therefore, the legal net flow is about 300,000 immigrants rather than 400,000. If so, the Census Bureau's current annual figure
of 400,000 immigrants could accommodate approximately 100,000 illegal
entrants a year without any change in the official population projections,
since these projections make no provisions for illegal immigration.

Studies by IEA (8,9) have led to the conclusion that a net annual flow of up
to 250,000 illegal entrants would probably make little difference to labor
force or GNP projections in this century. This judgment is based on the fact
that not all illegal immigrants enter the labor force, and the productivity of
those who do, because of language difficulties and other factors, is probably
well below that of native-born workers (9). However, if the flow amounts to
as many as 750,000 a year, as some estimates have it, then the consequences

for economic growth and energy demand would be significant, greatly off-setting the expected economic and demographic consequences of lower domestic fertility.

The number of illegal immigrants cannot be predicted with any accuracy. If legislation already pending before the Congress is enacted into law and enforced, the illegal entry problem could be reduced to small consequence, as it is in many European countries.

Labor Force

Labor force estimates to 2000 are not sensitive to population differences resulting from various assumptions about future FR, since it is not until about 1993 that labor force numbers would be affected. The decline in the rate of labor force growth occurs at the end of the entry of the postwar "baby boom" generation into the job market. This decline in growth begins by the late 1970s and accelerates thereafter. Labor force numbers could be larger if there is an acceleration in the participation rate, which defines the proportion of the population 16 years and older employed or actively seeking employment. The participation rate has been rising slowly. In 1960, it was 59.2 percent; in 1976, it was 62.1 percent. Thus, in 1976, of 156.0×10^6 persons of labor-force age, 96.9×10^6 were either employed or seeking employment and 59.1×10^6 were neither employed nor actively looking for work.

This large number of persons of working age who are not in the labor force is accounted for because the labor force is defined to include persons as young as 16 years of age; hence millions of the labor force "eligibles" are in school. Among the older groups, some are ill or disabled, some (an increasing proportion) are retired, and others may want to work but, for one reason or another, are not actively seeking jobs. But the largest number of unemployed are housekeepers. Clearly then, the participation rate will never approach 100 percent. The numbers in the age group 16 and over who were not in the labor force of 1976 are shown in Table 1-3. When the American economy had a much larger agricultural sector, the minimum age for inclusion in the labor force was 14. Perhaps in the face of a near-universal requirement to remain in school to age 18, the minimum age should again be raised to conform to current social practice.

Participation rates by men are expected to continue to decline slowly as projected by experts at the Bureau of Labor Statistics (11). This downward trend is in large part a reflection of earlier retirement made possible by Social

Table 1-3.

Nonparticipants in the labor force (10)

Reason	Millions of persons	Percentage
Total[a]	59.1	100.0
Going to school	6.2	10.5
Ill or disabled	4.7	8.0
Keeping house	31.0	52.5
Retired	8.4	14.2
Other	3.6	6.1
Want a job but not looking	5.5	9.3

[a]Data do not add because of the method of estimation used.

Security and supplementary retirement benefits. On the other hand, women are increasing their participation in the labor market and this trend is also expected to continue. Currently, about 45 percent of women 16 years of age and older are in the labor market, compared with 42.8 percent in 1970. We have projected that the female participation rate will increase to 52 percent by 2000. Table 1-4 gives the details for the FR projection of 1.7 children per female.

The number of employed males covered by Social Security reached 93 percent of the work force in 1975, or near universal coverage (12). It is not surprising then that the participation rate of males in the work force declines sharply at age 65. In 1950, this rate was 43 percent; in 1975, it was 21 percent.

The same trend was reflected in the male group 62 to 64 years old, which became eligible for retirement at reduced benefits in 1961. In addition, those who continue to work after reaching eligible retirement age often do so on a part-time basis, since the government-imposed earnings test limits the amount of income that can be earned, up to age 72, if a full pension is to be drawn from Social Security.

Although the participation rate for men is relatively constant (Table 1-4), that for women is not. The trend of rising female participation is acknowledged, but how far the trend will continue is a matter of judgment since there are no absolute limits to much higher participation rates for women.

Since women can retire on moderately reduced Social Security pensions at

age 62 and since most wives are younger than their spouses, the trend to retirement at age 62 for women is a natural consequence of the Social Security
law and social custom. Although female participation rates begin their decline
very early (after age 24), the sharpest reductions take place after age 60. In
1975, whereas about 48 percent of women in the age group from 55 to 59
years old were employed, only about 33 percent of those between 60 and 64
years old were still in the labor force and only about 8 percent of those 65 and
over.

If, in the future, women's participation continues to decline after age 62
and essentially stops after age 65, over 70 percent of the women between 18
and 62 years would have to enter the labor force to raise the overall participation to 55 percent. This would close virtually all of the current gap between
male and female participation rates.

Is this likely? Not if the FR stays in the range of 1.7 to 1.9 children per female, although, of course, a major shift in social customs could bring about
such a change. Women with children, particularly small children, do not participate in the labor force as fully as those without children. For example, in
1974 only one-third of the women 25 to 45 years of age with children under
6 years of age were in the labor force. For these reasons we are inclined to the
conclusion that an overall female participation rate of about 52 percent by
2000 is more likely than a much higher one.

No one knows for certain what will be the impact on either male or female
participation rates of the recently enacted law raising mandatory retirement

Table 1-4.

Labor force estimates: numbers (x 10^6) and participation rates
(in percentages)

Year	Men		Women		Total	
	Rate	No.	Rate	No.	Rate	No.
1975	78.5	59	45	36	61	95
1980	78	62	47	41	62	103
1985	77	65	49	45	62.5	110
1990	76.5	67	51	49	63	116
1995	76	68	52	51	63	119
2000	76	71	52	53	63.5	124

from 65 to 70 years of age. We do not believe it will alter current retirement patterns appreciably, since the very high rates of inflation of the past decade have had no measurable effect.

Hours Worked

Part-time employment estimates for the future reflect the long-term declining trend in hours worked, both for agricultural workers and for those in the private nonagricultural sector. Average hours worked per week in the years 1955 through 1968 declined 0.48 percent annually. What this has meant over the 25-year period from 1948 to 1973 (we have chosen 1973 as an end date here to avoid the decline caused by the depression that followed) is a drop in the total private nonagricultural workweek from an average of 40 hours in 1948 to 37.1 hours in 1973. Our projections of the future workweek anticipate a moderately slower further decline, at an annual rate of 0.40 percent. The slower rate of decline is in part a reflection of the expected slowdown in the rate of entry of women into the labor force. Experts at the Bureau of Labor Statistics noted that, over the period 1965 through 1968, there was a strong association between increasing participation rates of women, their frequency of part-time employment, and the average hours worked per week [(1), p. 3].

Future Unemployment Rate

National unemployment hit a post-World War II peak of 8.9 percent of the civilian labor force in the spring of 1975 and improved only moderately through 1977 to about 7 percent. The question is whether or not basic economic factors are at work that will maintain the level of unemployment at about 7 percent.

In the years since the close of World War II unemployment has averaged only slightly below 5 percent. It fell to 3 percent or less only in the years of virtually full employment during the Korean War, 1952 and 1953. Since the end of fighting in Vietnam, it has averaged well above 5 percent.

Some economists point to the changing composition of the labor force as an explanation of the current higher unemployment levels [(13), p. 52]. Teenagers now account for nearly 10 percent of the labor force; unemployment among teen-agers typically is quite high, in part a reflection of their relative lack of skills and experience. Moreover, adult women now make up a greater proportion of the total labor force, accounting for 36 percent of the total in 1976, a new high; unemployment among women who are not heads of households traditionally has been above that of male heads of households. The increased proportion of teen-age and adult women workers in the total labor force then is almost certainly one explanation for the recent high unemployment.

Another factor contributing to higher unemployment has been the rate of growth of the labor force, a consequence of the "baby boom" after World War II. Additions to the labor force are now close to 2.0 percent annually; they will start declining in the late 1970s. By the 1990s, the rate of growth is expected to fall off to about 0.3 percent annually in the absence of substantial immigration. Under these conditions, the average age (and experience level) of the labor force will increase and the annual absorption ratio of new entries will be much less than in the decade of the 1970s. Moreover, if our identification of causes of the high unemployment of the 1970s is correct, unemployment should fall to at least no higher than 5 percent—the long-run postwar average—and perhaps less.

There is a second major reason for believing that future unemployment is not likely to remain high. The Employment Act of 1946 established the objective of full employment, but no mandatory steps to its achievement were included in the act. However, it now seems likely that, under legislation supplementing the Humphrey-Hawkins Act, Congress will require the creation of employment in the public sector if unemployment continues to be high. The nation will thereby reach the level of "full employment" (14). Hence, we tend to view the present levels of joblessness, and the high rates of unemployment in the current business cycle, as a short-term phenomenon and have maintained the average postwar rate of 5 percent in our future projections.

Productivity Trends

The decline in productivity growth is a serious development, since, together with labor force growth, it determines the GNP estimate and is the main source of increasing living standards. Table 1-5 summarizes the statistics [(13), p. 48].

In commenting on the abrupt and sustained decrease in productivity, the report of the Congressional Budget Office stated [(13), p. 48]:

Apart from statistical problems, it is difficult even to identify, let alone quantify, the major causes of the slowdown in productivity growth. The most that can be done is to list some of the contributing factors together with broad judgments as to their probable importance.

Table 1-5.

Productivity growth in the private economy
(in percentage points)
(All figures exclude estimated variations in productivity due to
short-run output fluctuations.)

| | | Growth in output per worker due to | |
| | Average annual growth in output | More capital | Other |
Time period[a]	per worker	per worker	factors
1950-1955	3.2	1.2	2.0
1955-1960	2.7	0.7	2.0
1960-1965	2.7	0.7	2.0
1965-1970	2.4	0.9	1.5
1970-1975	1.0	0.4	0.6
1975-1977[b]	2.2	0.2	2.0

[a]Fourth quarter to fourth quarter.

[b]Forecast, from Peter Clark, Council of Economic Advisers.

The data for 1975 through 1976 showed improvement, but productivity
gains began to decelerate again in 1977 (15). Reasons advanced for the productivity change are many and are summarized below.

Capital per Worker
Although there are statistical problems in measuring capital stock, the slowdown in investment per worker is attributed primarily to the accelerated
growth of the labor force rather than to a slowdown in capital spending. The
civilian labor force grew 24 percent from 1965 to 1975 compared with 15
percent from 1955 to 1965. See Table 1-6 for data on the growth of capital
stock [(2), p. 50].

If the slowdown in investment per worker has been the most important
factor contributing to the productivity decline, as many believe, then the impact of this factor will be lessened as the drop in birth rates is reflected in
smaller relative and actual additions to the labor force, and productivity gains
should recover significantly.

Table 1-6.

Growth of the private capital stock, 1950 to 1977
(in percentage points)

| Time period[a] | Nonresidential fixed investment as a percentage of the GNP | Annual rate of growth | |
		Private effective capital stock[b]	Capital per worker
1950-1955	9.1	4.5	3.6
1955-1960	9.1	3.1	2.1
1960-1965	9.2	3.2	2.2
1965-1970	10.4	4.3	2.6
1970-1975	10.1	3.3	1.6
1975-1977[c]	9.5	2.5	1.0

[a]Fourth quarter to fourth quarter.

[b]Includes nonresidential plant and equipment and excludes investment in pollution abatement.

[c]Forecast [(13), p. 50; (16)].

The future decline in the number of 18-year-olds (Table 1-7) is based on projections of the Census Bureau Series II; past data are actual (17). The table illustrates the shift taking place in a prime age of entry into the labor force.

If IEA's estimates of the range of U.S. population growth (FR = 1.7 or 1.9 children per female) are reasonably correct, the two population figures (246×10^6 and 254×10^6) would give rise to only small differences (less than 1 percent) in the size of the labor force in 2000.

Investment in Pollution Abatement
Such investment is not considered as a contribution to increased productivity in our data, since its product is reflected in improved health and safety rather than in output measured in dollars. Perhaps the most dramatic impact of health and safety laws on productivity has occurred in the coal-mining industry, where the daily output per miner has been more than cut in half since

Table 1-7.

Changing numbers of 18-year-olds

Years	Change over 1 year
1970-1971	+95,000
1975-1976	+9,000
1980-1981	−65,000
1985-1986	−81,000
1990-1991	−207,000

1969. This decline, although dramatic for the coal industry itself, is statistically small compared to the total employment (perhaps 210,000 miners versus 88.4 million employed workers in the economy).

A recent analysis by Edward F. Denison (18) attributes much of the slower rate of economic growth to increased expenditures for government-mandated pollution and safety requirements. However, the ratio of investments in pollution abatement to total investments will almost certainly ease with time, since the greatest burden is for capital outlays necessary to upgrade existing public and private facilities, such as municipal water treatment plants. For example, direct industry capital costs for water cleanup, as estimated by the National Commission on Water Quality, would be $80 billion through 1983; thereafter, the annual operating costs would be $12 billion (19).

Capital Replacement Requirements

The growth of capital stock means that, over time, a larger proportion of the gross investment is needed to replace obsolete or worn-out plants and equipment [(13), p. 51]. Although there may have been some acceleration in plant and equipment replacement in recent years due to higher energy prices, this element cannot be isolated from industrial investment as a whole. Data on manufacturers' evaluation of the adequacy of plant and equipment facilities relative to prospective operations have been trending up since 1967, which would suggest that plant expansion rates have been adequate to meet anticipated demand (20).

Investment Trends in General

The availability of capital for new investment is a key variable for economic

growth, and the rate of introduction of new technology is an important determinant of productivity increases. Apart from investment questions raised
in the above paragraphs, there are now doubts that the rate of investment is
adequate. In the United States, the annual sum expended on fixed business
investment in the period from 1965 through 1970 was 10.4 percent of the
GNP (21).

According to the 1976 annual report of the Council of Economic Advisers
(CEA) (21), this ratio needs to increase to 12 percent of the GNP in order to
meet the goal of full employment, to reduce energy dependence, and to conform to environmental standards. Business investment as a share of the GNP
is much higher in some of the European countries and in Japan than in the
United States. Even though business investment in the United States expanded about 8 percent annually in 1976 and 1977, at the end of 1977 it had
regained only about three-quarters of the ground lost in the recession of 1974
through 1975 and growth has been slower than in business cycles in the earlier post-World War II period. Since the expansion of capacity has been relatively slow, the CEA has characterized the investment performance of the last
4 years as disturbing (22).

Theoretically, the supply of investment funds will match the demand for
capital. In practice, however, this outcome may not be achieved. Interest
rates may not always be free to perform their incentive and rationing func-
tions, and markets may not always clear, as the periodic shortage of funds for
housing frequently demonstrates.

To summarize, we note the following important trends affecting growth in
investment as a share of the GNP:

1. A continuing trend in investment toward the consumption-oriented service industries, many of which have a lower rate of energy use per worker
than manufacturing industries.

2. A growing share of investment going into replacement outlays as opposed
to new investment.

3. An upward trend of capital investment needed to produce a given level of
output as a result of new environmental standards.

4. A spectacular growth in transfer expenditures (assistance payments, pensions, welfare) by the federal government for which it receives no offsetting
goods or services.

5. Unprecedented federal, state, and local deficits which contribute to inflation and, in periods of high demand by the private sector, compete for financial resources with the private sector.

6. Future energy costs higher than those assumed in most current projections as investment per kilowatt-hour for nuclear-powered and coal-fired electricity-generating plants grows in response to higher environmental standards, and as more marginal sources of conventional fossil fuels are tapped. Current estimates of capital requirements for a given output from new energy sources are at least two to three times higher than estimates made only 5 years ago.

7. A continued reduction of "real" depreciation allowances and of the supply of investment funds as a consequence of continued inflation.

Low Profits
Since low profits reduce the funds available for business investment, the fact that profits, as a percentage of the GNP, fell in the period from 1970 through 1975 has been identified as a cause of lower investment. During periods of high inflation, depreciation allowances for tax purposes tend to fall well behind replacement costs, thus making replacement more costly to the firm.

Shifts in the Composition of the Labor Force
As noted above, since 1966 the composition of the labor force has shifted toward women and teen-agers, groups with relatively little work experience. If wages are a reasonable yardstick of their productivity (and wages may not be), then an increase in their share of the labor force would tend to lower the overall productivity growth. One observer estimates that wage differences and the composition of the labor force cannot account for more than a fifth of the decrease in the growth of factor productivity [(13), p. 52].

Other causes for reduced productivity growth have been advanced—inadequate expenditures on research and development (hardly a factor in the short run), higher energy costs, equal opportunity rules and other governmental regulations, and the flattening out of the growth in educational attainment of the work force (again a small causal factor at best).

We are not convinced that productivity growth has moved permanently to a lower track. The most plausible explanations are the sharp decline in the growth of capital per new worker as a consequence of the absorption of the "baby boom" population growth into the labor market and the diversion of investment funds to meet pollution and safety requirements. Productivity gains showed a rising trend from 1889 to 1965 (23). On the basis of our present knowledge, the assumption that a permanent productivity shift took place around 1965 is difficult to validate, although not an impossibility.

Hence, in our high growth projection (FR 1.9) in Table 1-8 we assume a full
recovery of productivity by 1985; in our low-growth projection (FR 1.7) we
assume that productivity recovers to about 92 percent of the achievement
from 1948 through 1955, to 2.6 percent a year (24).

Projections of the GNP

We need to carry out one additional calculation, the estimate of hours
worked per year. This number, multiplied by an estimate of productivity
growth (footnote e in Table 1-8), yields projections of the GNP.

The estimate of the hours worked in 2000 is based on our projection of (i)
the population 16 years and over, (ii) the anticipated participation rate of
each age group, (iii) an assumption that 5 percent of the labor force will be
unemployed, and (iv) a modest decline of 0.4 percent per year in the average
hours worked. (The derivation of all these factors has been explained above.)
The labor force participation rate is assumed to increase from 61 percent in
1975 to 63.5 percent by 2000, as a result of a continuation of the increased
participation of women. The hours worked by all employed (including part-
time workers, entrepreneurs, and the military) is reduced from the 1975
average of 35.7 hours a week to 32.3 hours for 2000 (see Table 1-8, which
includes these projections as well as those for the GNP). The total hours
worked in 2000 are computed as follows:

$$H = (P_w \times R) \, (1 - u) \, [(H75) \, (1 - r)^n]$$

where H is the total hours worked, P_w is the population of working age (16
years and over), R is the participation rate, u is the unemployment rate,
$H75$ is the hours worked per person in 1975, r is the annual reduction in the
hours worked, and n is the number of years from 1975. For the year 2000,
this calculation takes the form

$$197{,}406 \times 10^6 = (195{,}000 \times 10^3 \times 0.635)(1 - 0.05) \, [(1855)(1 - 0.004)^{25}]$$

$$197{,}406 \times 10^6 = (117{,}364 \times 10^3) \times 1678$$

The development of the estimates of the GNP is shown in Table 1-8. The estimate of the growth in the GNP per hour worked is based in part on the
stable 2.8 percent annual increase in productivity between 1948 and 1965
and the subsequent, much less stable increase of 1.6 percent per year for
1965 through 1975. Over a 25-year period, a 2.8 percent annual growth re-

Table 1-8.

Population, labor force, and the GNP

Year	Population (x 10^6)[a]		Labor force (x 10^6)[b]	
	FR 1.7	FR 1.9	FR 1.7	FR 1.9
1975	213	213	95	95
1985	229	231	111	111
2000	246	254	124	125

Year	Employed labor force (x 10^6)[c]		Hours worked (x 10^9)[d]	
	FR 1.7	FR 1.9	FR 1.7	FR 1.9
1975	87	87	161	161
1985	105	105	192	192
2000	118	119	198	200

Year	GNP per hour[e] ($1972)		GNP ($1972 x 10^9)	
	FR 1.7	FR 1.9	FR 1.7	FR 1.9
1975	7.39	7.39	$1,190	$1,190
1985	9.01	9.01	1,730	1,730
2000	13.24	13.24	2,620	2,648

[a]Data for FR 1.7 children per female are from (4) and correspond to Series III. Data for FR 1.9 children per female are an IEA calculation. Fertility rates shown are for the average number of live births per female at the completion of her childbearing.

[b]The labor force is made up of persons 16 years and over. Participation rates for all series are identical—1985, 62.5 percent; 2000, 63.5 percent.

[c]We assume a 5 percent unemployment.

[d]Assumes an annual 0.4 percent decline in the hours worked; the reduction is from 35.7 hours a week in 1975 to 32.3 hours in 2000.

[e]Assumes an FR of 1.7 and FR 1.9 children per female and a 1.8 percent growth in productivity in 1975-1980; 2.2 percent, 1980-1985; and 2.6 percent, 1985-2000. The annual growth of the GNP averages 3.8 percent, 1975-1985, and 2.8 percent, 1985-2000.

sults in a doubling of the GNP, whereas a 1.6 percent annual growth results in
an increase of less than 50 percent. The choice of the productivity growth
rate is, therefore, crucial to the results and overshadows all other variables
in the estimates. This is why it has been treated above in some detail.

Strictly interpreted, the hours worked record the hours paid for, not neces-
sarily the hours worked. One of the reasons for the recorded decline in productivity over the past decade may be a growing gap between hours paid for
and hours worked, as has been noted in interviews by Arthur Burns. We are
not aware of any statistically verifiable way to make this correction and have
not attempted to do so.

In the remainder of this chapter we will be concerned with the quantifica-
tion of key intermediate factors as shown in Figure 1-1—households, commercial space, and automobiles—which are essential to estimating quantities of
energy required to run the economy.

Households

Households are an intermediate factor from which one can calculate energy
demands for the household sector and the commercial sector. The number
and types of households used in this study are based on projections of the
Census Bureau to 1990 and IEA trend extrapolations between 1990 and
2000. Table 1-9 shows these estimates for various FR assumptions.

Because virtually all households are headed by adults, it is not until the mid-
1990s that the impact of a higher FR is even statistically perceptible. Variations in the consumption estimates of energy by households to 2000, therefore, are due largely to life-style differences (including conservation effects)
rather than to differences in the number of households.

In our projections it is anticipated that the trend to more single-person
households will probably peak by 1995. The trend of a long-term decline in
the average family size is expected to decrease the average area of house size,
although in many cases this decline is offset by the expansionary impact of
greater affluence, which would increase the area of the average house (25).
The trend to single-person households (about 23 percent of the total in 2000)
suggests lower numbers of single-family dwellings and more multifamily units.
Multifamily units, which comprised 31 percent of the total in 1970, are expected to increase to 40 percent by 1985. Multifamily units contain about
two-thirds of the area of new single-family homes. In addition to the impact
of smaller families and the trend to single-person households, higher land,

Table 1-9.

Estimated numbers of households, 1975 to 2000 (x 10^6)

Year	Households per adult[a]	Total households		Family households	
		FR 1.7	FR 1.9	FR 1.7	FR 1.9
1975	0.53	72	72	56	56
1985	0.55	87	87	66	66
2000	0.57	102	102	76	76

	Primary individual units[b]		Single-person units (subtotal of primary)	
	FR 1.7	FR 1.9	FR 1.7	FR 1.9
1975	16	16	14	14
1985	21	21	19	19
2000	26	26	23	23

[a]An adult is defined as a person 21 years of age or older.

[b]Primary individual households are either single-person households or multi-person households whose members are not related by blood or marriage.

construction, and energy costs make multifamily units more economical. A further space-decreasing trend is the growing use of mobile homes, which have about 600 square feet of living space as compared to about 1500 square feet in single-family homes. Mobile homes are often poorly insulated. We would conclude that changes in household composition should tend to decrease energy requirements for space heating and cooling, although there are increasing requirements for air-conditioning. We have not attempted to compute any "net savings."

Commercial Space

The commercial space estimates are a surrogate for a group of service industries which include wholesale and retail trade facilities, schools, government facilities, hospitals and nursing homes, and hotels. Not all service industries are included; both transportation and household services, two of the largest categories, are omitted because their energy requirements are calculated separately in this study. IEA has researched the growth trends of the service industries in detail (26). The projections for commercial space are based on this analysis. Analysis of economic activity in the postwar years does not bear out the impression that the service sector has been growing relative to the rest of the economy, whether we measure services as a share of output or as a source of income (see Table 1-10). Studies by the National Bureau of Economic Research (27) and the Bureau of Economic Analysis (28) support the conclusion that services have not grown as a proportion of income in the United States; international comparisons by the Organization for Economic Cooperation and Development (OECD) (29) and Simon Kuznets (30) have found no correlation between service production and the level of economic development.

Productivity gains in the service sector have consistently lagged behind those in the goods-producing sector, and a partial consequence has been the growth in service employment. The consequence is that the price of services has risen nearly 20 percent more than the price of goods over the past 25 years. If services are to grow more rapidly in the future so that their share of the gross domestic product (GDP) grows relative to goods (the GDP excludes the contribution of international activities to the total output), the productivity performance of services must improve. This is possible but not assured.

Our calculation for the projected growth of commercial space roughly parallels the GNP growth to 2000. This permits commercial requirements per household to increase 25 percent, from the present value of 350 square feet per household. This factor is then multiplied by the number of households to arrive at a total, 25.2×10^9 square feet. For 2000, we predict 437.5 square feet per household and 44.6×10^9 square feet total.

Since service industries are less energy-intensive than goods industries, a long-term secular shift to services could save about 1.4 quads of energy by 2000 [(25), p. 30]. We have not reduced our demand calculations for this shift, since it is not certain that the shift will take place, and, even if an allowance for it were made, the impact on the total energy demand would be small.

Automobiles

Automobile estimates are another key intermediate factor to projections of energy demands. There is uncertainty about the future trend in the use of the

Table 1-10.
Services as a share of income, measured as a proportion of output and final demand, 1947-1975
(share of the GDP in constant $1972)

Year	GDP	Energy	Owner-occupied housing[a]	Services			Nonservices		
				Private[b]	Public[c]	All	Goods[d]	Transportation[e]	All
Output									
1947	100.0	4.2	3.5	37.8	14.3	52.1	34.1	6.0	40.1
1958	100.0	4.4	5.8	36.8	14.8	51.6	33.9	4.2	38.1
1963	100.0	4.4	6.1	36.8	14.1	50.9	34.4	4.1	38.5
1967	100.0	4.5	6.0	37.1	14.1	51.2	34.2	4.1	38.3
1970	100.0	4.6	6.5	38.7	14.0	52.7	32.0	4.1	36.1
1972	100.0	4.4	6.3	38.5	13.2	51.7	33.6	3.9	37.5
1975	100.0	4.4	6.8	39.8	13.7	53.5	31.5	3.9	35.4
Final demand									
1947	100.0	2.9	3.5	35.9	12.5	48.4	39.1	6.2	45.3
1958	100.0	3.1	5.8	34.1	13.5	47.6	40.5	2.9	43.4
1963	100.0	3.1	6.1	34.3	12.8	47.1	41.5	2.3	43.8

1967	100.0	3.3	6.0	35.4	13.0	48.4	40.1	2.2	42.3
1970	100.0	3.6	6.5	35.4	12.9	48.3	39.6	2.1	41.7
1972	100.0	3.6	6.3	34.6	11.7	46.3	41.9	1.6	43.5
1975	100.0	3.9	6.8	35.9	12.2	48.1	39.7	1.6	41.3

Sources: Department of Commerce national income and product accounts, all years. Uses of income, 1972-1975, from unpublished INFORUM estimates, January 1976; price deflators from INFORUM data, for uses of income all years. Basic INFORUM model data are in Clopper Almon, Jr., et al, 1985: *Interindustry Forecasts for the American Economy*, D. C. Heath and Co., Lexington, Massachusetts, 1974. Uses of income, 1947-1970, from Department of Commerce input-output tables and unpublished working papers of the Bureau of Economic Analysis.

[a]Excludes the imputation for food consumed on the farm; space rent only.

[b]Excludes owner-occupied housing.

[c]General government plus government enterprise less public energy utilities.

[d]Agriculture, mining, and manufacturing.

[e]Excludes all owner-operated vehicles and public transportation.

private automobile. At the lower projection in Table 1-11 of 127 x 10^6 cars, the assumption is that the automobile market is near saturation and that public transportation will be increasingly important as the operating costs of private cars, including the escalation of fuel prices in the face of growing scarcity, make public transportation more attractive.

Those who believe that the higher projection in Table 1-11 of 152 x 10^6 automobiles is more likely cite the shift in population to the sunbelt states, which will require more driving and more cars than are necessary, for instance in built-up urban areas. Moreover, the increasing participation of women in the labor force could lead to a growing demand for private transportation for job-connected requirements. In 1975, there were 67 cars per 100 persons 16 years or older. This ratio is decreased to 65 per 100 persons by 1985 in the lower (101-quad) projection and remains constant thereafter to the year 2000. In 1975, there was approximately one automobile for every two persons (men, women, and children) in the nation. The lower projection predicts that there will be one car per 1.93 persons by 2000 (see Table 1-11).

The 126-quad projection is based on a continuation of the trend from 1964 to 1972 toward greater use of automobiles by licensed drivers.

In the next chapter of the study we establish quantitative relationships between energy demand and economic growth. We examine the energy required to operate the service and process equipment in the various sectors of the U.S. economy.

Table 1-11.

Estimated range of automobiles, 1975-2000
(The age groups 16 and over are calculated in the three fertility rates by the Census Bureau. The IEA low projection uses 1.7 children per female and the high projection uses 1.9 children per female).

Year	101-quad projection		126-quad projection	
	Number per 16 and over	Number of cars (x 10^6)	Number per 16 and over	Number of cars (x 10^6)
1975	0.67	104	0.67	105
1985	0.65	115	0.71	125
2000	0.65	127	0.77	152

On the basis of these energy demand relationships, we estimate future demands through the year 2000 (i) in light of the factors governing economic growth and energy consumption discussed in this chapter and (ii) expected improvements in the efficiencies of the various energy-use processes calculated in Chapter 2.

REFERENCES

1. E. L. Allen *et al., U.S. Energy and Economic Growth, 1975-2010,* Publication ORAU/IEA-76-7, Institute for Energy Analysis, Oak Ridge Associated Universities, Oak Ridge, Tennessee, 1976; C. E. Whittle *et al., Economic and Environmental Implications of a U.S. Nuclear Moratorium, 1985-2010,* Publication ORAU/IEA-76-4, Institute for Energy Analysis, Oak Ridge Associated Universities, Oak Ridge, Tennessee, 1976.

2. See, for example, C. T. Bowman and T. H. Morlan, "Revised Projections of the U.S. Economy to 1980 and 1985," Department of Labor, *Monthly Labor Review,* Washington, D.C., March 1976.

3. Edward F. Denison, *Accounting for United States Economic Growth, 1929-1969,* Brookings Institution, Washington, D.C., 1974.

4. U.S. Bureau of the Census, *Current Population Reports,* Series P-25, No. 704, "Projections of the Population of the United States: 1977 to 2050," U.S. Government Printing Office, Washington, D.C., July 1977.

5. Most of these data are taken from U.S. Bureau of the Census, *Population Characteristics,* Series P-20, No. 288, "Fertility History and Prospects of American Women: June 1975," U.S. Government Printing Office, Washington, D.C., 1976.

6. Charles F. Westoff, "Marriage and Fertility in the Developed Countries," *Scientific American,* vol. 239, No. 6, pp. 51-57.

7. U.S. Bureau of the Census, *Current Population Reports,* Series P-20, No. 307, "Population Profile of the United States: 1976," U.S. Government Printing Office, Washington, D.C., 1977, p. 6.

8. Institute for Energy Analysis, *Illegal Immigration and Future Economic Growth,* Oak Ridge Associated Universities, Oak Ridge, Tennessee, October 1976. See also Appendix A of this study.

9. Institute for Energy Analysis, *Analysis of Economic Growth Parameters,* Oak Ridge Associated Universities, Oak Ridge, Tennessee, May 1977, Appendix A.

10. U.S. Department of Labor, Office of Information, Release USDL 76-1012, July 12, 1976.

11. Howard Fullerton, "New Labor Force Projections to 1990," *Monthly Labor Review,* U.S. Department of Labor, Washington, D.C., December 1976.

12. *Economic Report of the President,* Washington, D.C., January 1976, p. 114.

13. Congressional Budget Office, Congress of the United States, *Sustaining a Balanced Expansion,* Washington, D.C., August 3, 1976.

14. The 4 percent level of unemployment as constituting "full employment" was defined by the Council of Economic Advisers early in the 1960s. See *Economic Report of the President,* "The Full Employment-Unemployment Rate," Washington, D.C., January 1977, pp. 48-51.

15. *Business Week,* "Productivity Is a Worry Again," August 22, 1977, pp. 22-23.

16. Part of the forecast for 1975 through 1977 is from Peter Clark, "Capital Formation and the Recent Productivity Slowdown," paper presented at the annual meeting of the American Economic Association, December 30, 1977.

17. U.S. Bureau of the Census, *Current Population Reports,* Series P-25, No. 601, "Projections of the Population of the United States: 1975 to 2050," U.S. Government Printing Office, Washington, D.C., 1975.

18. Edward F. Denison, "Effects of Selected Changes in the Institutional and Human Environment upon Output per Unit of Input," *Survey of Current Business,* Department of Commerce, Washington, D.C., vol. 58, No. 1, January 1978, pp. 21-44.

19. National Commission on Water Quality, *Issues and Findings,* Washington, D.C., November 1975, pp. I-15 through I-19. Expenditures for 1972 through 1976 are given in *Survey of Current Business,* Department of Commerce, Washington, D.C., vol. 58, No. 2, February 1978.

20. "Plant and Equipment Expenditures: 1977 Programs Revised," *Survey of Current Business,* Department of Commerce, Washington, D.C., vol. 57, No. 9, September 1977, p. 21.

21. *Economic Report of the President,* Washington, D.C., 1976, p. 44.

22. *Economic Report of the President,* Washington, D.C., 1978, p. 67.

23. Charles L. Schultze, *National Income Analysis,* Prentice-Hall, Englewood Cliffs, New Jersey, 1967, p. 121.

24. The 2.6 percent productivity estimate is identical to that of the Bureau of Labor Statistics for 1980 through 1985, although the Bureau of Labor Statistics uses a narrower definition of productivity. Ronald E. Kutscher, "Revised BLS Projections to 1980 and 1985: An Overview," *Monthly Labor Review*, Department of Labor, Washington, D.C., March 1976, p. 4.

25. Department of Housing and Urban Development, *Series Data Handbook, A Supplement to FHA Trends*, RR25, Washington, D.C., undated. In 1948, the floor area of new single-family homes averaged 972 square feet; by 1977, it had increased to 1313 square feet.

26. Robert W. Gilmer, Institute for Energy Analysis, *Services and Energy in U.S. Economic Growth*, Washington, D.C., 1977 (unpublished).

27. V. R. Fuchs, *The Service Economy*, Columbia University Press for the National Bureau of Economic Research, New York, 1968.

28. Edward Denison, "The Shift to Services and the Rate of Productivity Change," *Survey of Current Business*, Department of Commerce, Washington, D.C., vol. 53, No.10, October 1973, p. 21.

29. D. W. Blader, D. D. Johnston, W. Marzeaski, *Service Activities in Developing Countries*, OECD, Paris, 1974.

30. Simon Kuznets, *Economic Growth of Nations*, Harvard University Press, Cambridge, Massachusetts, 1971.

2

**FACTORS THAT INFLUENCE
ENERGY DEMAND AND
ECONOMIC GROWTH**

Introduction

In the preceding chapter we addressed the main determinants of economic
growth–population, labor force, and productivity. In this chapter we put for-
ward a view of future energy demand based on projections of several key determinants–conservation, government restrictions in energy use, saturation of
energy-consuming devices, and the impact of an aging population–which influence energy consumption.

This discussion deals with a wide range of issues, some on the "hard" and
some on the "soft" side of the social science spectrum. Several of the factors
considered lend themselves to reasonably rigorous analysis and to conclusions
which can be stated with a degree of confidence; others permit only judgmental consideration and tentative conclusions. We are sensitive to the perils of
forecasts and projections over several decades when so many determining factors, including individual whims and societal caprices, are unknown. On such
a time scale, it is well to acknowledge that a good "guess" is about as close an
approximation to the ultimate truth as may be possible from the present van-
tage point. However, the method used here is not different from that used by
those who have developed qualitative rationales for other energy-demand
scenarios [(1), Chapter 3].

The issues discussed here have significant implications for those responsible
for long-range energy planning. The discussion touches on sensitive aspects of

overall national energy policy which, in the past, have often been predicated
on a higher demand for energy than this analysis would indicate. Our approach has been to deal with what is *likely* to happen on the basis of currently available information, not with what *ought* to happen.

We have attempted to demonstrate a link between economic growth and
energy demand. This link depends on important economic, technical, and
social considerations. In the end, this relationship is vitally important in
establishing future energy requirements. The analysis establishing the specific
relationship between economic growth and energy demand is developed later
in this report.

Historical Energy-to-GNP Ratio

A crude correlation between energy consumption (E) and the gross national
product (GNP) can be established if one examines historical data (2). However, the ratio of energy to GNP has not been constant, as indicated by
Figures 2-1 and 2-2 (3). Most estimates (4) forecast that the ratio of energy
use to GNP will continue to decline. Figure 2-2, showing log E versus log
GNP, shows at least four periods of strong correlation between the two
variables. The long-term downward trend in the ratio of energy to GNP ob-
served from 1947 to 1967 reversed in 1967 and climbed until 1970. Since
1971 the ratio has continued downward, as shown in Table 2-1 (5).

What follows is a qualitative discussion of the key factors that affect the

Table 2-1.

Energy and the GNP

Year	Gross energy input ($\times 10^{15}$ Btu)	GNP ($\times 10^9$) ($ 1972)	Energy/GNP ratio ($\times 10^3$)
1947	33.0	$ 468.3	70.5
1950	34.0	533.5	63.7
1955	39.7	654.8	60.6
1960	44.6	736.8	60.5
1965	53.3	925.9	57.6
1970	67.1	1,075.3	62.4
1974	73.1	1,210.7	60.4
1975	71.1	1,186.4	59.9

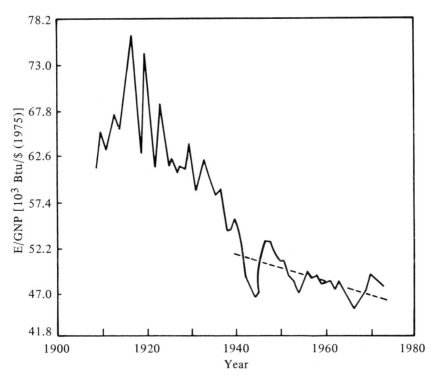

Figure 2-1.
Ratio of U.S. energy consumption to GNP. The dashed line is the least-
squares trend line for 1940 through 1970

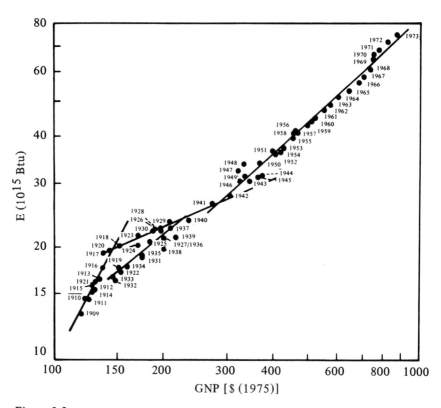

Figure 2-2.
Trends in U.S. energy consumption and GNP

energy/GNP ratio. That, in turn, permits an examination of the relationship between the growth of the GNP and growth in energy demands. Some of these factors, such as the conservation potential (the potential energy savings by various possible conservation steps) and losses in conversion, can be quantified. Others, such as the impact of changing life-styles, are more difficult to measure; consequently, the findings tend to be largely judgmental.

Conservation

Conservation (defined to include increases in energy-use efficiencies) appears to be the largest single variable that must be estimated in moving from growth in GNP to future growth in energy demand. As noted above, a decline in the energy/GNP ratio occurred during the postwar years, and that decline is continuing. The striking characteristic of this changing energy use has been the fact that, as a result of technical improvements and better energy management, the postwar savings took place in a period when energy was becoming less expensive relative to other commodities.

Energy prices fell relative to all other prices between 1947 and 1970. Since 1970, the price trends have reversed. The cost of energy, which once amounted to a trivial part of individual, corporate, and institutional budgets, has recently become more significant [(5), pp. 114-116]. Average fuel prices in the United States increased over 45 percent during 1974 and 1975. Although price increases have been more gradual recently, forecasts by IEA are for energy prices to increase more rapidly than the general rate of inflation. One can reasonably expect that, after a delay to allow for amortization of durable energy-consuming equipment, price pressures will bring about new and significant energy conservation measures even though domestic price controls are still in effect. Mandatory price regulations on petroleum, set up under the Energy Policy and Conservation Act of 1975, are due to expire in May 1979. They may be extended by the President, as may also be the case for natural gas, but the expectation is that domestic fuel prices will trend closer to world prices by 1980 and will equal them by 1985. In the longer-term future, conservation will be increasingly a cost-induced rather than an ethical or moral exercise; as such, it will have an important effect on total energy demand. Important government-mandated conservation measures are noted below under "Government Intervention."

Energy-saving technologies now exist which could be incorporated into the U.S. energy system to raise our energy productivity and sustain a reasonable

economic growth. Table 2-2 summarizes the results of several basic studies (1, 6-8) on the potential for conservation in the U.S. energy system. The implementation of a list of modest conservation technologies is proposed to
reduce the total U.S. energy requirements per unit of output, and to shift the
fuel demands from oil and gas to electricity and the direct use of coal. Heat
from the cogeneration of electricity can also play a significant role. The in-
corporation of the suggested new technologies can be timed to coincide with
the normal retirement of capital stock in transportation, housing, commerce,
and manufacturing, without placing an undue burden on investments which
otherwise would be required under a more austere and controlled program.

In addition to the studies listed in Table 2-2, we have examined other
studies on savings that may be realized in specific sectors. These include
studies by the National Petroleum Council (9), Dow Chemical Company (10),
Oak Ridge National Laboratory (11-13), and the Council on Environmental
Quality (14). In this study we have made conservative assumptions regarding
the effect of potential improvements in end-use efficiencies. Each of our assumptions is discussed below when we consider the transportation, residential and commercial, and industrial energy sectors.

A study sponsored by the Federal Energy Administration (15) compares
energy consumption between West Germany and the United States by sector
on a per capita basis. The results are summarized in Table 2-3.

These comparisons between West Germany and the United States illustrate
the potential energy savings possible in the United States if energy prices rise
high enough and if feasible conservation goals are followed. West Germany
uses only one-half as much energy per capita as the United States. Total
energy use in West Germany, relative to national income, was only about two-
thirds that of the United States; the ratio of energy use to national income
for residential use was 48 percent, and the ratio for industrial use was 58
percent.

Some of the differences in per capita energy uses in transportation for both
passengers and freight can be explained by the great differences in population
density; the West German population density is about ten times that of the
United States. Another reason for the differences is most likely the higher
fuel prices in West Germany, which result in the use of more efficient autos
and greater use of public transportation systems.

Some of the differences in the residential and commercial sectors can be explained in terms of the life-styles that are characteristic of the two societies.
Other differences are due to building design and insulation, methods of

Table 2-2.
Potential energy savings

Reference	Base demand in 2000 (quads)	Saving potential (x 10^{15} Btu)			Total (quads)	Savings by 2000 (%)
		Transportation	Residential and commercial	Industrial		
(6)	192	5.0	23.5	23.0	51.5	26.8
(7) (Scenario 0 to Scenario 1)	165	7.4	15.3	19.4	42.1	25.5
(1) (Base case to technical fix)[a]	187	13.7	15.2	33.8	62.7	33.5
This study	149.3	5.2	7.0	11.2	23.4	15.7
(8) (Potential savings for 1973 only)	75 (1973 demand)	9.1	11.6	10.4	31.1	41.5 (percent demand)

[a]The "base case" is the projection of the industrial growth rate for energy for the period 1950 through 1970, 3.4 percent per year. The "technical fix" scenario lowers this rate to 1.9 percent per year by conservation measures.

Table 2-3.

West Germany per capita energy use as a percentage
of U.S. per capita energy use (15)

Sector	Percent
Transportation total	27
	—
Road transport	29
Air transport	20
Railroads	75
Inland and coastal shipping	33
Military	13
Residential total	48
	—
Space heating	67
Heating water	37
Cooking	60
Air-conditioning and clothes drying	1
Other	18
Commercial	56
Industrial	58
	—
Total	49

heating water, and varying levels of appliance saturation. One of the interesting contrasts in energy use between West Germany and the United States can be seen by examining the energy requirements per unit of output in some of the industrial sectors, as shown in Tables 2-4 and 2-5.

One finding of the Energy Policy Project of the Ford Foundation [(1), pp. 45-79] was that, through available conservation techniques, U.S. energy demand in the year 2000 could be reduced from an estimated 187 quads (the Ford Foundation "high" scenario) to 124 quads (the "technical fix" scenario), a reduction of one-third. According to the Ford Foundation study, such a reduction would have little effect on the growth of per capita income.

In terms of total energy requirements needed to sustain economic growth and employment, our low-projection totals (101 quads by 2000) are not very different from those given in the Ford Foundation zero energy growth case. However, our low projection was reached not by imposing life-style changes such as the substitution of mass transit and bicycles for personal automobiles on the United States (as was the case in the Ford Foundation study). It was reached by calculating technically feasible energy requirements per unit of output which could be carried out in the future and by our judgment that future economic growth rates will be less than those used in the Ford Foundation study.

Conservation possibilities extend to all forms of energy consumption (16, 17). Industrial consumption, the largest single energy-using sector, can benefit from improved boiler design and heat recovery processes. The manufacture of lighter-weight automobiles and service trucks with more efficient engines could double the average miles per gallon (mpg) for the operational fleet by the year 2000, compared to present efficiencies. Household and commercial consumption could improve with better building design and insulation, improved heat pumps and the solar-assisted annual cycle energy system (ACES) (18), and the retrofitting of existing buildings.

All these changes will take time. The number of new dwelling units built each year, for example, amounts to only about 3 percent of the existing inventory. Retrofitting existing buildings at the rate of 1 to 2 percent each year increases the improvements to a 4 to 5 percent rate. Changing over the automobile inventory is a decade-long task. As we shall show below, what can be accomplished in the longer run without straining the limits of existing technology is impressive.

Table 2-4.

Industrial energy use for 1972: the United States and West Germany (15)

Industry	Total (x 10^{15} Btu)		10^3 Btu/$ of shipments		10^3 Btu/employee	
	United States	West Germany	United States	West Germany	United States	West Germany
Primary metals	5.65	1.21	97	78	5.0	2.0
(iron/steel/aluminum)	(3.63)	(0.98)	(151)	(111)	(7.7)	(3.1)
Chemicals	4.10	0.73	72	41	4.9	1.3
Petroleum and coal products	3.20	0.38	112	56	23.0	9.1
(petroleum refining)	(3.12)	(0.36)	(120)	(55)	(30.9)	(9.8)
Paper	2.93	0.17	104	39	4.6	0.9
Stone, clay, and glass	1.61	0.49	75	55	2.6	1.2
Food processing	1.36	0.17	12	8	0.9	0.4
Other manufacturing	4.20	0.77	9	7	0.3	0.1
Other industries (mining and nonindustrial)	4.43	0.77	—	—	—	—
Total	27.48	4.69	35	25	1.4	0.6

Table 2-5.

Energy/output ratios of four industries (15)

Industry	10^6 Btu per short ton	
	United States	West Germany
Iron	45.0	27.8
Steel	27.3	18.6
Paper	46.6	26.6
Petroleum products	4.9	3.3

Government Intervention

Since the Organization of Petroleum Exporting Countries (OPEC) oil embargo in late 1973, the federal government has taken a number of steps to
encourage, or require, energy conservation. The Federal Energy Management
Program, established by presidential order in June 1973, required a 15 percent cut in anticipated energy consumption by federal government agencies
in fiscal year 1974. In fact, energy consumption was cut by 24 percent during
the subsequent 12-month period; about half of these savings were attributed
to reduced use of jet fuel by the Department of Defense. Other savings followed changes in lighting standards and heating and cooling specifications.
This program has been continued (Table 2-6).

   A second early conservation measure was the introduction of the 55 mile-
per-hour speed limit on a national basis. Related federal measures were di-
rected at voluntary savings of gasoline through the formation of car pools,
more frequent tune-ups, and use of smaller cars. In the consumer goods field,
voluntary recycling of returnable bottles received federal support.

   The federal government has moved in a variety of directions to lower energy
consumption for specific users. In the transportation field, Public Law 94-163
(December 22, 1975) established average mpg standards for manufacturers,
beginning with the 1978 model year and becoming progressively more stringent through 1985 (Table 2-6). The 1985 standard, 27.5 mpg, is an improvement of more than 50 percent over the 1975 value. The 1977 models incorporated more efficient engines and lighter bodies, anticipating the 1978
requirement. The same law requires disclosure of energy consumption and

Table 2-6.

Conservation and government intervention measures

Action taken	Enforcement standards	Expected results
Households		
Public Law 94-385, August 14, 1976 Development and implementation of performance standards for new residential buildings whether federally financed or not. States to adjust building codes to comply. Standards to be developed in 3 years, implemented within 1 year.	No federal loans except on structures in compliance.	Federal Energy Administration (FEA) states proper insulation can save 20 to 30 percent of heating and cooling bills.
Public Law 94-163, December 22, 1975 Administrator, Environmental Protection Agency (EPA), sets minimum energy improvement targets for: (1) refrigerators and refrigerator-freezers (2) freezers (3) dishwashers (4) clothes dryers (5) water heaters (6) room air-conditioners (7) home heating equipment (8) television sets (9) kitchen ranges and ovens (10) clothes washers that are manufactured in 1980 to be 20 percent more improved than in 1972	All products must be labeled to indicate the energy consumed.	By 1980: maximum possible percentage improvement of aggregate energy efficiency, but not less than 20 percent.

Administrator, EPA, by December 22, 1976, prescribes improvement target for: (11) humidifiers and dehumidifiers (12) central air-conditioners (13) furnaces manufactured in 1980	All products must be labeled to indicate the energy consumed.	By 1980: percentage improvement of aggregate energy efficiency to be set by Administrator.
H. R. 6860, August 27, 1975 Credit of 30 percent of qualified insulation (insulation that meets federal standards) and other expenditures for energy-conserving equipment which does not exceed $750. Credit of 40 percent of qualified expenditures for solar energy equipment plus qualified expenditures for geothermal equipment which does not exceed $1000. Credit of 25 percent with respect to any sum paid thereafter which exceeds $1000 but not $7400.	Income tax allowances.	Designed to improve heating and cooling systems by 20 to 30 percent.
Qualified solar energy equipment: qualified heat and geothermal energy equipment, pump equipment and qualified wind-related energy equipment. Credit of 20 percent for qualified heat pump expenditures instead of 40 percent as above. Credit of 12.5 percent instead of 25 percent as above for expenditures over $1000 and less than $7400.	Income tax allowances.	
Public Law 94-385, August 14, 1976 Grants to states and to Indian tribes to provide for weatherization of dwelling units, especially where elderly or handicapped	Judiciary Board if there is a complaint.	Provide better insulation to reduce house heating and cooling energy consumption by 20 to 30 percent.

Table 2-6.
Conservation and government intervention measures (Cont'd)

Action taken	Enforcement standards	Expected results
Households		
low-income persons reside and where there is a low-income head of the household. Grants not to exceed $55 million ending September 30, 1977, and $65 million ending September 30, 1978. Standards to be developed in 60 days and after comments on regulations; they must be implemented in 30 days (Title III). Title IV provides for financial assistance to state conservation programs designed to save energy in existing buildings.		
Commercial		
H. R. 6860 12 percent tax credit for insulation for pre-1980 property; 22 percent tax credit for solar energy equipment and geo-thermal energy equipment installed in pre-1980 property. Public Law 94-385, August 14, 1976 (Same as for households under Titles III and IV)	Measure not yet enacted by Congress. Would be an addition to investment tax credit.	Designed to make it more attractive for commercial buildings to achieve heating and cooling savings of 20 to 30 percent.

Public Law 94-385,
August 14, 1976
(Same as for households)

Industrial

H. R. 6860
12 percent tax credit for insulation for
pre-1980 property; 22 percent tax credit
for solar energy equipment and geothermal energy equipment installed in pre-
1980 property.

Would be an addition to investment tax
credit.

Public Law 94-385,
August 14, 1976
(Same as for households under Titles III
and IV)

Designed to improve the heating and
cooling by 20 to 30 percent.

Transportation

Public Law 94-163,
December 22, 1975
Average fuel economy for new passenger
cars, beginning in the 1978 model year,
not less than:

Liable for $5 for each tenth of a mpg by
which average fuel economy standard ex-
ceeds the average mpg multiplied by the
total number of passenger cars manu-
factured by manufacturer.

By 1985 average fuel economy for
passenger autos should be 27.5 mpg.

Table 2-6.
Conservation and government intervention measures (Cont'd)

Action taken	Enforcement standards	Expected results
Transportation		
Average mpg		
1978 18.0		
1979 19.0		
1980 20.0		
1981 Determined by the Secretary of Transportation		
1982 Determined by the Secretary of Transportation		
1983 Determined by the Secretary of Transportation		
1984 Determined by the Secretary of Transportation		
1985 onward 27.5		
Secretary of Transportation will prescribe average fuel economy standards for light trucks, beginning in 1980.	Liable for $5 for each tenth of a mpg by which average fuel economy standard exceeds average mpg multiplied by the number of light trucks manufactured by manufacturer.	Result in steady progress toward standard in 1985.

operating costs by manufacturers so that consumers will be aware of the
energy costs associated with each prospective car purchase. The Secretary
of Transportation will also prescribe fuel economies for light trucks beginning in 1980.

For household appliances, energy labels showing the annual operating costs
will also be required, so that consumers will be aware of the relative energy
costs of competing brands of the same appliance. By 1980, energy-use improvement for a wide range of consumer products must be at least 20 percent, compared to the figure for 1972 (Table 2-6).

Public Law 94-385, Title III, the "Energy Conservation Standards for New
Building Acts of 1976," requires within 3 years the development of federal
performance standards for new residential and commercial buildings to prevent energy waste (Table 2-6). It will operate through state and local building
codes and through federal financial assistance programs covering commercial
and residential buildings. The term "commercial building" means any building other than a residence, including buildings for industrial and public purposes. Federal buildings must meet or exceed the construction standards
finally promulgated.

Public Law 94-385, Title IV, the "Energy Conservation in Existing Buildings
Act of 1976," covers energy conservation in dwelling units, commercial
buildings, and industrial plants. It provides grants to states and Indian tribes to
be used for the weatherization of dwellings, especially for low-income and
handicapped persons. The authorization for appropriations for weatherization ranges from $55 million in the first year to $80 million in the third year.
Part B provides for the development of state energy conservation plans and
for federal financial assistance for the implementation of such plans. Appropriations authorizations are $25 million for fiscal year 1977, increasing to
$40 million by fiscal year 1979. The Federal Energy Administration (FEA)
announced in July 1976 that more than 6.7 million single-family homeowners
in ten states would have the opportunity to participate in a joint federal/state
energy-saving program. A variety of approaches will be used in the early
stages. Home heating and cooling of single-family dwellings account for 12
percent of the total national energy use; according to FEA experts, nearly 80
percent of the nation's 47 million single-family dwellings are inadequately insulated (19). Estimated savings from simple corrective action, consisting primarily of insulation and caulking, are stated to run between 15 and 30 percent.

Programs likely to have far-reaching effects, if enacted by Congress, are
President Carter's proposals for tax credits to homeowners and tax credits to

commercial and industrial enterprises for insulation and other energy-conserving expenditures. Proposed tax credits would also cover the installation of solar energy equipment.

Saturation of Energy Uses

One consideration frequently missing in historical projections of energy consumption is that of saturated markets. In the past, the idea that demand for almost any consumer-durable good would slacken off was rarely considered; with an inexorable rise in the overall standard of living and in population, it seemed unlikely that anything approaching market saturation could take place. Yet this has already occurred for electric lighting in the home. Telephones are another example: in 1941, only about 40 percent of American homes had telephones; by 1975, more than 94 percent had telephones and many had two or more (20). Automobiles are yet another example: there are more automobiles registered in the United States than licensed drivers.

In a July 1976 news release, the U.S. Department of Transportation announced that the number of licensed drivers reached 814 per 1000 persons of driving age in 1975 (21). In many states (Kansas, Montana, Nebraska, Nevada, New Mexico, Ohio, West Virginia, and Wyoming) licensed drivers were approaching a maximum, over 900 per 1000 persons of driving age. This ratio cannot reach 1:1 for a variety of reasons, one of which is the number of handicapped persons. States where the ratio is lowest (New York and the District of Columbia) are high in the per capita income listings, but the availability of urban transportation has led to a different life-style.

In part, recent increases in the numbers of privately owned automobiles reflect the fact that the current period—up to 1980—is one in which the labor force is increasing rapidly. This growth, now running about 2 percent a year, is expected to decline to less than 0.5 percent in the late 1990s, as explained in Chapter 1.

The future growth of automobiles in service will be geared much more closely to the annual increase in the size of the driving population. Since the number of new drivers (that is, those people reaching 16 years of age) has now stabilized and will soon slacken significantly [the Census Bureau Series III population projection estimates a decline of 670,000 in 1985 as compared to 1975 (22)], a slackening in the rate of growth of the automobile inventory is likely. This is especially significant because of the important role the growth in the number of automobiles has played in the growth of energy demand. Automobile transportation uses took 25 percent of the total petro-

leum consumption in 1975 (Btu basis). Although gasoline consumption grew
at an annual rate of about 5 percent over the past decade, price increases and
the improvement in the number of miles per gallon should reduce this growth
and bring about an absolute decline in gasoline consumption in the early
1980s.

A second area of likely saturation, which is important although not accompanied by the same energy-saving potential as passenger cars, is household
appliances. The post-World War II years have seen the approach of saturation
for most of these, for example, refrigerators, cooking ranges, space heaters,
and water heaters. A variety of surveys report that virtually every American
household, "poor" or "well-off," is equipped with stoves, refrigerators, and
television sets; almost half of the lower-middle income households have air-
conditioners and clothes dryers; almost two-thirds of the "poor" households
have washing machines [(1), pp. 113–122] . The Ford Foundation's Energy
Policy Project study (1) placed the time of saturation for most currently
available household appliances at about 1985.

The future growth in household energy consumption will be more closely
tied to the increase in the total number of housing units, without the additional consumption which has typically accompanied the expansion of appliances for existing housing units in the past. Moreover, in view of the
increasing share of total households that are single-person households, we
would expect a moderate shift to commercial suppliers of household services
(laundry, fast food services, and others) rather than a full complement of
energy-consuming appliances in each household. New appliances, such as
trash compacting units, may gain popularity, but the market seems close to
saturation for most major basic energy-consuming appliances.

Energy Demands and an Aging Population

The age composition of the U.S. population has shifted upward so that there
are now fewer young persons and proportionately more old persons in the
population. This effect is dramatically confirmed by the closing of elemen-
tary schools across the nation. This "aging" of the population is a consequence of a lower fertility rate plus improved health care. One important consequence is the rapid increase in the number of single-person households. This
development, discussed in Chapter 1, is primarily a reflection of the growing
affluence of retired persons.

The flow of federal funds to the elderly has increased dramatically in the past 10 yeras. In fiscal year 1969, Social Security payments totaled $25 billion and there were 22.5 x 10^6 recipients. Congress voted in 1972 to increase benefits and to adjust Social Security payments annually for inflation starting in 1975. It is thus not surprising that the number of elderly families estimated to be below the poverty level dropped from 28 to 14 percent in the decade ending with 1975. In fiscal year 1979, Social Security outlays will be about $90 billion to about 30 x 10^6 recipients (see Table 2-7). By 1985, the number of Social Security recipients is expected to rise to 33 x 10^6.

A second consequence of the aging of the population is the drop in the proportion of preschool and school age children. This decrease is so large that the relative size of the population in the prime labor force has expanded. Also, the participation of women of childbearing age in the labor force tends to increase as the fertility rate drops. In 1974, 37 percent of women aged 20 to 24 who had preschool children were in the labor force, compared to a total participation rate of 63 percent for all women 20 to 24 years old (23).

What are the implications for energy consumption of the shifts in age composition? Table 2-8, based on Census Bureau Series III, gives estimates of the numbers of persons expected to be in each of four age brackets that are particularly pertinent to our study from the standpoint of energy consumption per capita. These groups are defined somewhat arbitrarily to give a clearer picture of shifts in the relative sizes of those age groups that are significant double users of energy and those that are not (24). Age groups have been chosen to reflect the status of the great majority in each group (see Table 2-8). Specifically, preschool children and retired persons almost certainly consume less energy per capita than persons who are in school or working. The latter two groups not only consume energy as members of a household but also either at school or at the place of employment. This means a second per capita energy requirement for space and water heating, air-conditioning, and lighting in schools, offices, or factories (space heating is the largest energy end use in the household sector). Both the in-school and working age groups are major daily claimants on transportation services, which account for the second largest energy end use. In 1975, 21 x 10^6 children were transported to and from school by buses. The average number of miles driven by persons 60 years and older is less than by persons of labor force age (25).

From Table 2-8 it can be seen that the combined school age and work age groups, as defined above, accounted for 79.4 percent of the total population in 1975; by 2000, these same two groups are expected to represent 79.3 percent, which, given the margin of error around any social extrapolation, can

Table 2-7.
Federal spending for the elderly, 1969-1979.

Outlay	Fiscal 1969 Outlays (billions of dollars)	Fiscal 1969 Beneficiaries (x 10^6)	Fiscal 1979 Outlays (billions of dollars)	Fiscal 1979 Beneficiaries (x 10^6)
Social Security	$ 24.7	22.2	$ 90.1	30.0
Medicare (hospitals)[a]	4.8	19.7	21.0	26.6
		(4.4)[b]		(6.1)[b]
Medicare (doctors)[a]	1.8	18.8	9.1	26.4
		(9.0)[b]		(16.4)[b]
Medicaid	0.9	3.2	4.2	3.5
Civil Service retirement	2.4	0.9	12.2	1.6
Military retirement[a]	2.4	0.7	10.1	1.3
Railroad retirement	1.5	1.2	4.2	1.2
Supplemental Security Income	1.2	2.0	1.8	1.6
Total	$ 39.7		$152.7	
All federal outlays	$184.5		$500.2	
Spending on elderly as percent of total outlays	21.5		30.5	

[a]Includes disability payments.
[b]Number actually receiving payments.
Source: *National Journal*, February 18, 1978.

Table 2-8.

Population projections by age groups
(in millions of people) (22)

Age group	1975		2000[a]	
	Number	Percentage of total	Number	Percentage of total
All ages	213	100.0	246	100.0
Preschool (under 5)	16	7.5	14	5.7
School age (5 to 18)	55	25.8	45	18.3
Work age (19 to 61)	114	53.6	150	61.0
Retirement age (62 and over)	28	13.1	37	15.0
Median age	(28.8)	—	(37.0)	—

[a]Based on Bureau of the Census Series III.

be considered an identity. The dramatic shifts in age brackets, then, are not in low-energy users, the preschool and retired groups, but between those of school age and working age, both of which are major energy consumers.

We conclude that the energy impact of the changing age composition of the population in 2000 is toward higher per capita consumption of energy, largely because the active work force consumes more energy than school children or the retired consume. Industrial processes are much more energy-intensive than any other activity. For example, industry in 1975 consumed about three times as much energy as commercial activity and nearly twice as much as households. A rough calculation is that the projected shifting age composition to the year 2000, in itself, would increase the total energy consumption by 2 percent.

REFERENCES

1. Ford Foundation Energy Policy Project, *A Time to Choose: America's Energy Future,* Ballinger, Cambridge, Massachusetts, 1974.

2. Walter G. Dupree, Jr., and John S. Corsentino, "United States Energy Through the Year 2000 (Revised)," Bureau of Mines, U.S. Department of the Interior, Washington, D.C., 1975, p. 6.

3. Joseph D. Parent, "Some Comments on Energy Consumption and GNP," an unpublished paper prepared at the Institute for Gas Technology, Chicago, Illinois, May 1974.

4. Petroleum Industry Research Foundation, *The Outlook for World Oil into the 21st Century, with Emphasis on the Period to 1990* (EA-745, SOA-76-328), New York, 1978, Chapter 2.

5. Series taken from tables in *Economic Report of the President,* Council of Economic Advisers, Washington, D.C., 1976.

6. "The Nation's Energy Future," A Report to the President of the United States, submitted by Dixy Lee Ray, WASH-1281, U.S. Government Printing Office, Washington, D.C., December 1, 1973.

7. "Creating Energy Choices for the Future," The Plan, ERDA-48, U.S. Government Printing Office, Washington, D.C., June 30, 1975, vol. 1.

8. Hugh C. Wolfe, Series Editor, *Efficient Use of Energy,* Conference Proceedings No. 25, American Institute of Physics, New York, August 1975.

9. "Potential for Energy Conservation in the United States," Part I: 1974-1978 (September 10, 1974); Part II: 1979-1985 (August 6, 1975), National Petroleum Council, Washington, D.C.

10. James R. Burroughs, "The Technical Aspects of the Conservation of Energy for Industrial Processes," a report to the Federal Power Commission National Power Survey, Technical Advisory Committee on the Conservation of Energy, Position Paper No. 17, Dow Chemical Company, Midland, Michigan, May 1, 1973.

11. R. S. Carlsmith, Principal Investigator, "Energy Conservation Studies, Progress Report—December 31, 1974," Oak Ridge National Laboratory Report No. ORNL-NSF-EP-84, Oak Ridge, Tennessee, March 1975.

12. Eric Hirst, "Energy Use for Food in the United States," Oak Ridge National Laboratory Report No. ORNL-NSF-EP-57, Oak Ridge, Tennessee, October 1973.

13. T. D. Anderson, H. I. Bowers, R. H. Bryan, J. L. Delene, E. C. Hise, J. E. Jones, Jr., O. H. Klepper, S. A. Reed, I. Spiewack, "An Assessment of Industrial Energy Options Based on Coal and Nuclear Systems," Oak

Ridge National Laboratory Report No. ORNL-4995, Oak Ridge, Tennessee, July 1975.

14. "Energy Conservation in the Manufacturing Sector, 1954-1990," sponsored by the Council on Environmental Quality, contained in the Federal Energy Administration's Project Independence Blueprint Final Task Force Report 4118-00048, U.S. Government Printing Office, Washington, D.C., 1974.

15. Richard L. Goen and R. K. White, "Comparison Between West Germany and the United States," Stanford Research Institute Report No. SRI EGU 3519 (Federal Energy Administration Report No. FEA-D-75-590; National Technical Information Service Report No. PB-245 652), Stanford Research Institute, Menlo Park, California, June 1975.

16. Denis Hayes, "Energy: The Case for Conservation," Worldwatch Institute Paper No. 4, Washington, D.C., January 1976.

17. Bruce Hannon, "Energy Conservation and the Future," *Science,* vol. 189 (1975), pp. 95-102.

18. H. C. Fischer, J. C. Christian, E. C. Hise, A. S. Holman, A. J. Miller, W. R. Mixon, J. C. Moyers, E. A. Nephew, *The Annual Cycle Energy System: Initial Investigation,* Report ORNL/TM-5525, Oak Ridge National Laboratory, Oak Ridge, Tennessee, October 1976.

19. Federal Energy Agency, *Federal Energy News,* Washington, D.C., July 8, 1976.

20. Data furnished by the American Telephone and Telegraph Company (Helen Tomasini), New York, March 9, 1976.

21. U.S. Department of Transportation, Federal Highway Administration, *Drivers Licenses, 1975,* Washington, D.C., July 1976.

22. U.S. Bureau of the Census, *Current Population Reports,* Series P-25, No. 601, "Projections of the Population of the United States: 1975 to 2050," U.S. Government Printing Office, Washington, D.C., 1975.

23. Department of Labor, *Monthly Labor Review,* vol. 98, No. 11, November 1975, pp. 17-24.

24. School attendance continues beyond age 18 for those going on to college, but there is a sharp decrease in school attendance after the completion of secondary education. Similarly, there is some significant participation in the labor force after age 62, particularly by men. However, a secular decline is already appearing in the participation of those in the age group from 55 to 59. Participation beyond age 65 is relatively small and is expected to continue to decline. See Bureau of the Census, *Demo-*

graphic Aspects of Aging and the Older Population in the United States, Special Study Series P-223, No. 59, Washington, D.C., May 1976.

25. D. B. Shonka, A. S. Loebl, P. D. Patterson, *Transportation Energy Conservation Data Book,* Oak Ridge National Laboratory Report CONS/ 7405-1, Oak Ridge, Tennessee, 1977, p. 52. For sale by the U.S. Government Printing Office, Washington, D.C.

3

**TWO ENERGY DEMAND
SCENARIOS: 101 AND
126 QUADS**

Energy Demand

Energy analysts divide energy use into four broad economic sectors—household, commercial, industrial, and transportation. Estimates of the final energy demands in all IEA studies have begun with the population projections, then have moved to a consideration of the intermediate factors of households, commercial space, automobiles, and the gross national product (GNP), and then have gone on to an analysis of the categories of energy use for each economic sector. The specific energy demands obtained from an analysis of each sector are then summed to obtain total estimates of energy demand. Growth in the number of energy-consuming devices or units in each sector—the number of automobiles, the number of households, the amount of commercial floor space required—is tied to population growth estimates. Freight and air transport and the output of each manufacturing sector are primarily tied to economic growth (or GNP) estimates.

We have examined the amounts of energy currently needed to operate major energy-consuming devices or processes along with the potential for future energy conservation. The technical strategies we have used for conservation were based on the modest introduction of currently available technology, timed to coincide with normal replacement of capital stock. One obtains the total energy demand for each case by adding up the sector demands on the basis of specified energy use and economic activities.

Energy Demand in 1975

The year 1975 provides the base from which projections are made in both
scenarios. Hence, it is necessary to review briefly the energy supply and demand sectors for 1975. Each of the broad energy end-use sectors uses electricity, although the transportation sector consumes only a trivial amount.
The largest uses of electricity are for residential and commercial lighting and
appliances, and for industrial electric drive (electricity input to drive electric
motors) and lighting. Space heating and cooling, process steam and heating,
aluminum processing, and the iron and steel industries use the balance of the
electricity.

The household and commercial sectors in combination consumed about 35
percent of the total 1975 fuel input to the United States and depended primarily on supplies of electricity, natural gas, and fuel oil. The largest energy
use in these sectors is for space heating and cooling. Air-conditioning consumes only a small amount of energy compared to space heating, less than
one-fifth, in both households and commercial establishments. Forty percent
of the fuel for heating and cooling was obtained directly from natural gas,
40 percent directly from fuel oil, 18 percent from electricity generated from
various fuels, and 2 percent directly from the use of coal. Lighting and small
electric appliances and hot-water heaters and large appliances which use electricity, natural gas, and fuel oil constitute the balance in these sectors (see
Table 3-1).

The transportation sector (automobiles and the transport of goods) relies
primarily on liquid fuels and represented about 26 percent of the total fuel
input to the U.S. energy system in 1975. Approximately 60 percent of transport sector use (15.6 percent of the total) is for gasoline to power automobiles and light service trucks. Mandated improvements in average fuel economy for new passenger automobiles (Public Law 94-163, December 22, 1975)
range from 18.0 miles per gallon (mpg) in model year 1978 to 27.5 mpg in
model year 1985. The same law gives the Secretary of Transportation authority to prescribe average fuel economy standards for light trucks, beginning in
1980. These mandated changes will have the largest impact on the transportation sector energy use. The other transport categories—trucks, trains, buses,
and tractors, which use primarily diesel fuel; air transport, which uses primarily jet fuel; and ships, barges, and pipelines, which use diesel fuel and
natural gas—make up the balance of the transportation sector. All of these
transportation modes offer the potential for better fuel efficiencies (see
Table 3-1).

Table 3-1.

Summary of 1975 U.S. energy demands by source
(fuel inputs to sectors in 10¹⁵ Btu)

Use category	Coal	Oil	Gas	Electricity	Other fuels	Total
Households	—	3.8	4.9	7.1	—	15.8
Space heating and cooling		3.1	3.7	1.5		8.3
Lighting and small appliances				3.8		3.8
Water heating and large appliances		0.7	1.2	1.8		3.7
Commercial space	0.3	2.0	2.5	4.5	—	9.3
Space heating and cooling	0.3	1.9	2.1	0.9		5.2
Lighting and small appliances				3.3		3.3
Water heating and large appliances		0.1	0.4	0.3		0.8
Automobiles	—	8.3	—	—	—	8.3
Transport of goods and services	—	9.6	0.6	0.1	—	10.3
Service vehicles		2.8				2.8
Truck or rail or bus or tractor		3.5		0.1		3.6
Air transport		2.4				2.4

Table 3-1.
Summary of 1975 U.S. energy demands by source
(fuel inputs to sectors in 10^{15} Btu) (Numbers in parentheses are not included in totals.) (Cont'd)

Use category	Coal	Oil	Gas	Electricity	Other fuels	Total
Ship or barge or pipeline		0.9	0.6			1.5
Industrial processes	4.3	5.7	9.0	8.4	—	27.4
Process steam and heat	1.7	1.9	8.3			11.9
Iron and steel	2.4			0.4		2.8
Aluminum				0.8		0.8
Electric drive and lighting				7.2		7.2
Feedstocks	0.2	3.8	0.7			4.7
Electricity inputs	8.8	3.3	3.2	(20.1)	4.8	—
Totals (10^{15} Btu)	13.4	32.7	20.2	—	4.8	71.1
Percent of total	18.7	46.0	28.4	(28.3)	6.9	100.0

Source: Press release, Bureau of Mines, U.S. Department of Interior, March 1976. The final revision of 1975 demand, given in a Bureau of Mines press release dated March 14, 1977, totals 70.6 quads. Individual subsector adjustments have not been made in the above table, since they are not significant.

The industrial sector used about 39.5 percent of the total U.S. fuel input in 1975 (see Table 3-1). About one-sixth of this (or 6.6 percent of the total) was used for nonfuel purposes such as petrochemical feedstocks. The remaining 33 percent was used for process steam and heat; for electric drive and lighting; and for iron, steel, and aluminum processing. The largest single industrial use of energy is for process steam and heat (42 percent of the sector demand and 16.7 percent of the U.S. total). Most of this is for (i) petroleum refining, (ii) metal processing, (iii) the manufacture of chemicals and allied products, (iv) the pulp and paper industry, (v) food processing, and (vi) the production of stone, clay, and glass products. Over 80 percent of the total U.S. industrial energy use was concentrated in these six industries. Almost 80 percent of the fuel for process steam and heat came from oil and natural gas. The balance represents electricity and the direct use of coal. The largest user of coal, other than the electric utilities, was the iron and steel industry (1).

The 101-Quad Demand Scenario

In the 101-quad demand scenario for the year 2000 we assume a fertility rate of 1.7 children per female. We also assume the achievement of effective but not maximum conservation. We assume that more efficient technologies now in the process of development would contribute substantially to conservation by the year 2000. These technologies include (i) new building construction, with improved design and heat insulation standards, and with electric heat pump systems and a heat storage tank for heating and cooling; (ii) smaller and lighter-weight automobiles and service trucks with more efficient engines and transmissions, and containing less steel; and (iii) industrial boiler design and heat recovery processes in the various energy-intensive manufacturing industries with fuel shifted from oil and gas to the direct use of coal and nuclear heat or to electricity. We assume that the rate of introduction of these technologies will follow the normal retirement of extant capital stock, estimated in Table 3-2. An effective but achievable program of conservation is incorporated under the demand estimates given for household, commercial, transportation, and industrial uses.

Calculation of Household Energy Demands to 2000
Old Homes. We begin with the total number of houses in 1975, 72×10^6 (2), and a retirement rate of 2 percent a year. The stock of old houses remaining in the year 2000, then, is

Table 3-2.

Assumed average life for various devices

Energy-consuming device	Assumed average life (years)
Houses	50
Heating and cooling system (household)	25
Water heaters and large appliances	15
Electric lighting systems	25
Small appliances	5
Electric light bulbs	1
Commercial buildings	50
Heating and cooling system (commercial)	25
Water heaters and large appliances	15
Automobiles	10
Service trucks	8
Industrial boilers	15
Iron and steel furnaces	25
Aluminum smelters	25
Feedstock users (chemical and petroleum)	25

$$\text{Stock}_{2000} = 0.98^{25} (72 \times 10^6) = 43.4 \times 10^6$$

where 25 is the number of years until 2000. Factoring in a 2 percent retrofit rate per year gives an initial number of 1.44×10^6 retrofitted houses for 1975 and a declining number of retrofitted houses thereafter; the cumulative retrofits by 2000 is 29.4×10^6. A retrofit consists primarily of insulation measures which increase the energy efficiency of the house to 0.80 (3). The 1975 house energy intensity (energy consumption per house) was 118.1×10^6 Btu/unit. Thus, we calculate an energy intensity for 2000 based on insulation measures of

$$0.80 (118.1 \times 10^6) = 94.5 \times 10^6 \text{ Btu/unit}$$

The total energy demand is thus 2.5 quads on the assumption that 90 percent of the retrofitted houses (26.5×10^6) have insulation only, that is, (26.5×10^6) (118.6×10^6). The other 10 percent are assumed to have a heat pump installed, as well as the insulation retrofit. Thus, estimating an energy efficiency improvement to 0.65 for the heat pump, we calculate

$$(0.80)\,(0.65)\,(118.1 \times 10^6) = 61.4 \times 10^6 \text{ Btu/unit}$$

Applying this energy intensity to the remaining portion of retrofitted houses, we calculate an energy demand of 0.2 quad. Combining both groups (insulation only plus insulation and a heat pump) yields an energy demand of 2.7 quads of household consumption.

Of the surviving old houses in 2000 (43.4×10^6), 29.4×10^6 have been improved, leaving 14.0×10^6 unchanged. Using the 1975 energy intensity of 118.1×10^6 Btu/unit for unimproved houses yields an energy demand of 1.6 quads, that is, $(18.1 \times 10^6)\,(14.0 \times 10^6)$.

New Homes (Built after 1975). Our estimated house total for 2000 is 102×10^6; old homes (1975 or earlier) surviving in 2000 number 43.4×10^6, and new houses (built after 1975) total 58.6×10^6. We assume that all new houses are insulated and that 25 percent are equipped with a heat pump, so that our intensity calculation for these houses is

$$(0.25)\,(0.80)\,(0.65)\,(118.1 \times 10^6)\,(58.6 \times 10^6) = 0.9 \text{ quad}$$

For the 75 percent of new housing units that do not have a heat pump, energy demand is given by

$$(0.75)\,(0.80)\,(118.1 \times 10^6)\,(58.6 \times 10^6) = 4.2 \text{ quads}$$

This gives a new home total demand of 5.1 quads. Current (1977) production of heat pumps would permit about 25 percent of new houses to be equipped. However, an unknown percentage of heat pumps are being used for retrofit and for commercial applications.

Total Heating and Cooling. The household space heating and cooling demand estimate for 2000 becomes

New homes	5.1 quads
Old homes retrofitted	2.7 quads
Old homes unchanged	1.6 quads
Total	9.4 quads

Water Heating, Lighting, and Appliances. For these applications, we assume an efficiency factor of 0.90 (energy required per appliance) and a 1 percent annual increase in intensity (growth in the stock of appliances) due to rising affluence. This would allow for saturation of air-conditioning, trash compactors, and other new household appliances. This gives a total household demand of 12.4 quads.

A summary of the household energy demand projections is given in Table 3-3.

Commercial Energy Demand to 2000

In the commercial sector, much the same procedure is used as in the household sector but with the following differences:

1. In the commercial sector, our estimates of square feet of space are the key statistic.

2. Growth of the commercial sector is based on projections for the household sector. Between 1950 and 1975, service employment as a share of the civilian labor force grew from 45 to 61 percent. We anticipate that service employment will increase to 2000 but at a slower rate. Commercial space per household is assumed to increase by 25 percent to the year 2000 (see the section on "Commercial Space," Chapter 1), from 350 to 437.5 square feet.

3. Commercial buildings are already better insulated and designed than residential buildings. Our energy conservation estimates for commercial buildings are somewhat lower, at an efficiency of 0.90.

Old Commercial Establishments. Assuming a 50-year lifespan and a 2 percent retirement rate per year, we project old commercial footage in 2000 at 15.2×10^9 square feet. At a retrofit rate of 2 percent a year, we project retrofitted space in 2000 at 6.1×10^9 square feet.

The energy intensity in 1975 was 0.21×10^6 Btu/square foot. Assuming that all of the retrofits are insulated and that 10 percent are also retrofitted with heat pumps, we calculate a total heating and cooling demand of 1.1 quads, as follows:

(no heat pump) $(0.9) (6.1 \times 10^9) (0.90) (0.21 \times 10^6)$ = 1.0 quad
(with heat pump) $(0.1) (6.1 \times 10^9) (0.90) (0.65) (0.21 \times 10^6)$ = 0.1 quad

                                                      1.1 quads

The nonretrofitted area is $15.2 \times 10^9 - 6.1 \times 10^9 = 9.1 \times 10^9$ square feet. At the 1975 energy intensity factor of 0.21×10^6 Btu/square foot, this yields a total energy demand of 1.9 quads. Combining this total with the total for retrofits, we have 3.0 quads.

Table 3-3.

Household energy demand projections:
101-quad and 126-quad scenarios

Parameter	1975	2000, 101-quad	2000, 126-quad
Totals—energy inputs (x 10^{15} Btu)	16.2	21.8	29.1
Households (x 10^6 units)	72	102	102
Heating and cooling (x 10^{15} Btu)	8.5	9.4	13.00
New units (design factor)[a]	1.0	0.80	0.80
New units (equipment efficiency)[b]	1.0	0.65	0.65
Retrofits (insulation) (1%/year)	1.0	0.80	0.80
Retrofits (equipment efficiency)	1.0	0.65	0.65
Energy intensity (x 10^6 Btu/unit)	118.1	92.2	118.60
Water heating and large appliances (x 10^{15} Btu)	3.8	6.1	7.9
New units (design and efficiency)	1.0	0.90	0.90
Energy intensity (x 10^6 Btu/unit)	52.7	59.8	77.9
Electric lighting and small appliances (x 10^{15} Btu)	3.9	6.3	8.2
New units (design and efficiency)	1.0	0.90	0.90
Energy intensity (x 10^6 Btu/unit)	54.1	61.8	80.2

[a]Design factor efficiency improvements result from structural changes in buildings.

[b]Equipment efficiencies are, in part, federally mandated improvements in the hardware (refrigeration, for example).

New Commercial Establishments. Our projection of new commercial footage is 29.4×10^6 square feet. Assuming that one-quarter of this footage is fitted with a heat pump, we calculate a heating and cooling total for this sector of 5.1 quads, as follows:

(no heat pump) (0.75) (29.4×10^6) (0.90) (0.21×10^6) = 4.2 quads
(with heat pump) (0.25) (29.4×10^6) (0.90) (0.65) (0.21×10^6) = 0.9 quad
                                                                          —————
                                                                          5.1 quads

Water Heating, Lighting, and Appliances. For these applications we again assume an efficiency factor of 0.90. If appliance demand parallels the growth in commercial footage and factoring in the 1975 energy intensities, we project a total demand in 2000 of 6.7 quads, as follows:

water heating and large appliances:
   $(0.8/25.2 \times 10^6)$ (0.90) (44.6×10^6) = 1.3 quads
electric lighting and small appliances:
   $(3.4/25.2 \times 10^6)$ (0.90) (44.6×10^6) = 5.4 quads

where 25.2×10^6 is the number of square feet of commercial space in 1975, 44.6×10^6 is the number of square feet of commercial space in 2000, 0.8 quad is the energy demand for water heating and large appliances in 1975, and 3.4 quads is the energy demand for electric lighting and small appliances in 1975.

Structural Changes in Commercial Demand. Finally, we assume that the composition of the service sector will shift from more energy-intensive demands to less energy-intensive ones. Essentially, this reflects the impact of an aging population in the service sector—a shift away from schools to medical care and greater expenditures on commercial amusements. Listed in Table 3-4 are relationships between Btu's per dollar of output based on Bureau of Commerce data, which illustrate this impact on individual service sector categories. Consequently, we factor in a 0.5 percent annual decrease in energy intensity (because of the expected shift from more energy-intensive demands to less energy-intensive ones). Total demand is given by the following: 8.1 quads (heating and cooling) plus 6.7 quads (water heating, lighting, and appliances) equals 14.8 quads. The expected structural shift in service demand reduces the total commercial demand to 12.6 quads. The commercial energy demand projections are summarized in Table 3-5.

Table 3-4.

Comparison of energy requirements for service sectors (4)

Title	Btu/$1971 output
Private higher education	34,844
Private elementary and secondary schools	34,844
Other private education	32,723
Physicians	10,345
Private hospitals	26,196
Commercial amusements (including recreation)	18,718

Transportation Sector

The transportation sector is broken down into five categories: (i) automobiles, (ii) service trucks, (iii) truck and bus and rail freight, (iv) air transport, and (v) ship and barge and pipeline transport. Our projections of energy demand in this sector are based on estimates of end use and fuel efficiency.

We multiply the number of automobiles by the miles driven divided by the miles per gallon to calculate the gallons of fuel. We assume that the annual mileage per automobile will remain at the current 10,000 and fuel efficiency will increase to 20 miles per gallon (mpg) by 1985 and 27 mpg by 2000. These efficiencies are conservative by comparison with standards incorporated in recent energy legislation (Public Law 94-163). On the basis of these assumptions, fuel demand will decrease to 7.5 quads by 1985 and 6.1 quads by 2000.

With respect to service trucks, we assume that vehicle miles will grow at about the same rate as the GNP. Combining this assumption with fuel efficiencies of 14 mpg by 1985 and 18 mpg by 2000 gives a fuel demand of 1.5 quads in 1985 and 1.7 quads in 2000. Making this same assumption about the growth of truck, bus, and rail freight—in terms of ton-miles—but projecting a more modest fuel efficiency factor, we obtain the total fuel demands for this sector. Assuming that fuel demand per ton-mile can be reduced 10 percent by 1985 and 20 percent by 2000, totals of 4.9 and 6.9 quads are estimated, respectively.

Table 3-5.

Commercial energy demand projection:
101-quad and 126-quad scenarios

Parameter	1975	2000, 101-quad	2000, 126-quad
Totals—energy inputs (x 10^{15} Btu)	9.5	12.6[a]	16.2[b]
Commercial space (x 10^9 square feet)	25.2	44.6	44.6
Heating and cooling (x 10^{15} Btu)	5.3	8.1	8.4
New units (design factor)	1.0	0.90	0.90
New units (equipment efficiency)	1.0	0.65	0.65
Retrofits (insulation) (1%/year)	1.0	0.90	0.90
Retrofits (equipment efficiency)	1.0	0.65	0.65
Energy intensity (x 10^5 Btu/square foot)	2.11	1.82	1.88
Water heating and large appliances (x 10^{15} Btu)	0.8	1.3	1.5
New units (design and efficiency)	1.0	0.90	0.90
Energy intensity (10^{15} Btu/square foot)	0.32	0.29	0.33
Electric lighting and small appliances (x 10^{15} Btu)	3.4	5.4	6.2
New units (design and efficiency)	1.0	0.90	0.90
Energy intensity (x 10^5 Btu/square foot)	1.31	1.21	1.41

[a]Column adds to 14.8, but structural shift lowers total to 12.6 (see text).

[b]Totals may not add as a result of rounding.

For air transport and for ship and barge and pipeline transport, we assume fuel efficiency improvements of 10 percent by 1985 and 20 percent by 2000. Using the same growth assumptions as for service trucks, we estimate that growth will approximate that of the GNP. This gives a total of 2.4 quads in 1985 and 3.3 quads in 2000 for air transport and 1.5 quads in 1985 and 2.0 quads in 2000 for ships and barges and pipelines (see Table 3-6).

Industrial Demand

To forecast industrial energy needs, we estimated for each of the major industries a growth rate and changes in the amount of energy required to produce a given output. The industries examined in detail account for 85 percent of the energy used in manufacturing. They include (i) chemicals, (ii) primary metals, (iii) petroleum, (iv) stone, clay, and glass, (v) paper, (vi) food, and (vii) motor vehicles.

Estimates for other manufacturing, mining, and construction energy use were added. The energy used in agriculture to operate trucks and tractors was included in the transportation sector rather than in the industry sector. Each of these estimates was based on an examination of past trends and on a consideration of a variety of technological, economic, and demographic factors likely to affect the future.

The estimate of 44.4 quads for industrial energy demand in 2000 in this study (see Table 3-14) is below earlier estimates made by IEA (5) and is a result of the changing composition of industry. Petroleum and chemicals, the two industries that accounted for most of the increase in manufacturing energy consumption in the past 20 years, are forecast to grow much less rapidly in the next two decades. Furthermore, we often found that within a particular industry the more energy-intensive products are likely to grow less rapidly than other products, a result of higher energy prices. The energy input per dollar's worth of output in this study also decreased somewhat faster than in earlier studies by IEA (5), as new technology for saving industrial energy appeared more promising and the response to higher energy prices became more evident. For the period from 1954 to 1974, the average annual decline in the energy/output ratio was 1.6 percent. Between 1971 and 1974, when energy prices rose precipitously, the ratio was 5.2 percent. Although such a rate of decline cannot continue, the evidence does suggest that industry can and will become more energy-efficient.

The growth estimates for each industry are based on (i) past growth, largely 1954 to 1974; (ii) IEA estimates of future growth of the GNP, population,

Table 3-6.
Transportation energy demand projections: 101-quad scenario

Parameter		1975	1985	2000
Totals—energy inputs	(× 10^{15} Btu)	18.6	19.4	22.2[b]
Automobiles	(× 10^6 vehicles)	104	115	127
	(× 10^{11} vehicle-miles)	1.04	1.15	1.27
	(Miles per gallon)	14	20	27
	(× 10^{15} Btu)	9.8	7.5	6.1
Service trucks	(× 10^9 vehicle-miles) (%/year)[a]	99 (3.8)	144 (2.8)	217
	(Miles per gallon)	11	14	18
	(× 10^{15} Btu)	1.3	1.5	1.7
Truck and bus and rail freight	(× 10^9 ton-miles) (%/year)[a]	505 (3.8)	734 (2.8)	1111
	(Efficiency factor)	1.0	0.9	0.8
	(× 10^3 Btu/ton-mile)	7.1	6.8	5.9
	(× 10^{15} Btu)	3.6	5.0	6.9
Air transport	(Efficiency factor)	1.0	0.9	0.8
	(× 10^{15} Btu) (%/year growth)[a]	2.4 (3.8)	3.3 (2.8)	4.6
Ship and barge and pipeline transport	(Efficiency factor)	1.0	0.9	0.8
	(× 10^{15} Btu) (%/year growth)[a]	1.5 (3.8)	2.1 (2.8)	3.0

[a]Growth factors (in percent per year) are shown in parentheses for service and freight sectors for each time period.

[b]Totals may not add as a result of rounding.

and household formation; (iii) a review of estimates by others such as the
Office of Business Economics; and (iv) judgments on price and demand developments for some major products. Table 3-7 summarizes the growth projections for each industry.

Estimates of the energy/output ratios are based on (i) past trends, (ii) the
technological potential for reducing the ratio in specific industries, and (iii)
an examination of industrial energy use in other countries. Table 3-8 summarizes the projected changes in the energy/output ratio to 1985 and 2000.
Note the large decline in the years 1971 through 1974, when energy prices
rose rapidly.

These estimates are for net energy demand. In terms of gross energy, which
takes account of energy losses in the generation and transmission of electricity, the decline in the energy/output ratio is smaller. For the six major industries, we estimate that the net energy required to produce a dollar's worth
of output in 2000 will be 64 percent of that required in 1974. For gross energy we estimate about 70 percent.

The Six Major Industries
To compute net energy demand, we first project output. For example, in the
chemical industry, output in 1974 was valued at $81.4 billion. At the esti-
mated growth rate of 5.3 percent per year, output in 1985 would be $143.6
billion. The energy demand in 1974 was 2.9 quads. The energy/output ratio
is thus 2.9/81.4 = 0.036. Projecting an annual 2.1 percent decline in energy
consumption per dollar value of output to 1985, this ratio would drop to
0.029. Applying this new ratio to the 1985 output projection gives the 1985
energy demand projection, (0.029) (143.6) = 4.1 quads, where $143.6 billion
is the estimated dollar value of chemical output in 1985.

These six major industries (Table 3-8) account for 80 percent of the total
energy consumed by manufacturing industries. The remaining manufacturing
industries were treated separately as automotive manufacturing and other
manufacturing because growth in the automotive industry is expected to be
substantially smaller, at 3.0 percent to 1985 and 0.6 percent to 2000. For the
six major industries growth is projected at the GNP growth rate, 3.8 percent
to 1985 and 2.8 percent from 1985 to 2000. In either case the annual energy/
output ratio is projected to decline at 1.5 percent to 1985 and 2.0 percent to
2000.

Total energy demand in auto manufacturing in 1974 was 0.2 quad. The auto
output was 6.9×10^6 units. Projecting 2000 output at 10.4×10^6 units and

Table 3-7.

Annual estimated growth in production for the major manufacturing industries, 1985 and 2000;
SIC, standard industrial classification.

Industry	SIC Code	Average annual percentage growth		
		1954 to 1974	1974 to 1985	1985 to 2000
Chemicals and allied products	28	8.2	5.3	5.5
Primary metals	33	3.3	3.0	2.8
Petroleum and coal products	29	3.7	1.0	1.0
Stone, clay, and glass	32	3.6	2.5	2.5
Paper and allied products	26	5.0	4.5	3.5
Food and kindred products	20	3.6	3.2	3.2
Six industries listed above		4.5	3.4	3.5
Total manufacturing		4.5	3.6	3.0

Table 3-8.
Estimated annual percentage change in the energy/output ratio for the major manufacturing industries, 1985 and 2000

Industry	SIC Code	Average annual percentage change			
		1954 to 1974	1971 to 1974	1974 to 1985	1985 to 2000
Chemicals and allied products	28	-3.6	-5.0	-2.1	-2.8
Primary metals	33	-1.6	-5.1	-1.2	-1.0
Petroleum and coal products	29	+0.7	-4.7	-1.5	-1.5
Stone, clay, and glass	32	-3.7	-4.8	-3.5	-3.2
Paper and allied products	26	-1.3	-4.3	-1.7	-1.5
Food and kindred products	20	-2.6	-5.3	-2.5	-2.0
Six industries listed above		-1.3	-3.6	-1.5	-1.9
All manufacturing		-1.6	-5.2	-1.5	-2.0

factoring in the declining energy/output ratio, we project energy demand at
0.2 quad, implying little' or no change from 1974:

$$(10.4 \times 10^6)(0.2/6.9 \times 10^6)(0.985)^{11}(0.980)^{15} = 0.2 \text{ quad}$$

where 98.5 = 100 − 1.5 and 98.0 = 100 − 2.0 from Table 3-8. A similar technique applied to other manufacturing generates an energy demand projection
in 2000 of 3.2 quads.

When we add our demand projections for the six major industries, auto
manufacturing, and other manufacturing, we derive the 1985 and 2000 totals
shown in Tables 3-9 and 3-10.

Gross Energy Demand

Our procedure to determine the gross energy demand follows the IEA
method used to estimate future electrification [(5), pp. 97-99]. The increasing use of electricity is a consequence of increasing supply and price constraints on natural gas and petroleum, plus the improved electrical efficiency
made possible by the heat pump. This study projects total industrial electricity use to increase by 4.8 percent a year to 1985 and 3.8 percent thereafter to
2000. We modify this rate across industries to take account of projected developments. For example, as oil and gas become more expensive, the paper
industry may switch to the direct burning of coal. Hence, we discount the 4.8
percent electric growth rate in papermaking to account for this development.
But in the steel industry we augment this growth rate because of the greater
use of the electric furnace.

We begin with industrial use of electricity in 1974. A specific growth rate
is factored in to obtain a figure for the year 2000. This figure is net energy;
hence it is multiplied by 3 to obtain the gross energy needed to generate the
electricity used. Adding this figure to the direct fuel total (fuel bill minus
electrical bill) gives the gross energy demand. This procedure generates the
totals shown in Table 3-11.

Feedstocks and Captive Coke

The industrial energy data on which this study is based are the figures for purchased fuel and electricity for heat and power, as reported in (6). These data
do not include the feedstocks for the petroleum and the chemical industries

Table 3-9.

Estimates of output and net energy requirements for the six major industries, 1974, 1985, and 2000

Industry	Output (billions of $1974)			Purchased fuel and electricity (quads)		
	1974	1985	2000	1974	1985	2000
Chemicals	81.4	143.6	320.6	2.9	4.1	5.9
Primary metals	92.4	127.9	193.5	2.6	3.2	4.2
Petroleum and coal products	56.9	63.4	73.6	1.6	1.5	1.5
Stone, clay, and glass	26.7	35.0	50.7	1.3	1.2	1.1
Paper and allied products	39.8	64.6	108.2	1.3	2.2	2.7
Food and kindred products	156.7	221.7	355.5	1.0	1.0	1.2
Total	453.9	656.2	1102.2	10.7	13.2	16.6

Table 3-10.

Net manufacturing energy demand projections, 1974, 1985, and 2000
(quads)

Sector	1974	1985	2000
Six major industries	10.6	13.2	16.6
Motor vehicles	0.2	0.3	0.2
Other manufacturing	2.4	3.0	3.2
Total	13.2	16.5	20.0

Table 3-11.

Gross manufacturing energy consumption, 1974, 1985, and 2000
(quads)

Sector	1974	1985	2000
Total, six major industries	14.0	17.6	23.5
Motor vehicles	0.4	0.4	0.5
Other manufacturing	4.1	5.4	7.1
Total	18.5	23.4	31.1

or the captive coke (coke used to manufacture steel) in the primary metals in-dustry. To incorporate these additional factors, we project 1974 figures for-ward to 1985 and 2000 at the rate of growth that is estimated for the indus-tries that use them. The 1974 total is 4.9 quads. For 1985 and 2000 it is 6.0 and 8.5 quads, as shown in Table 3-12.

Mining
In 1972 the mining industry used about 2.3 quads of gross energy, distributed among the mining of the following: metals, anthracite coal, bituminous coal and lignite, oil and gas, and nonmetallic minerals. A detailed investigation of these processes generated estimates of production changes and energy/output changes to the year 2000, and these estimates form the basis of IEA projec-tions for 2000 (Table 3-13). The projected total is 3.1 quads.

Construction
A projection of energy demand in the construction industry based on house-hold growth to 2000 yields a total of 1.7 quads. Since the figure is so small, no refinement to estimate a more exact energy demand figure has been at-tempted.

Table 3-12.
Energy consumed by feedstocks and captive coke, 1974, 1985, and 2000 (quads)

Sector	1974		1985		2000	
Feedstocks	2.4		2.9		4.4	
Chemicals		1.6		2.2		3.7
Petroleum		1.0		0.7		0.7
Captive coke	2.6		3.1		4.1	
Total	5.0		6.0		8.5	

Table 3-13.

Energy use in mining by industry, 1972 and 2000 (quads)

Mining sector	1972 Energy demand	Production, 2000/1972	Energy/output, 2000/1972	2000 Energy demand
Metals	0.25	1.5	1.1	0.42
Anthracite coal	0.01	2.0	0.92	0.02
Bituminous coal and lignite	0.14	2.0	0.92	0.25
Oil and gas	1.53	0.9	1.3	1.79
Nonmetallic minerals	0.37	2.0	0.8	0.58
Total	2.30			3.05

Total Gross Industrial Energy Demand

Combining the totals for manufacturing, feedstocks and captive coke, mining, and construction, we obtain the total industrial demand figures given in Table 3-14.

Total Energy Demand

Our demand projections for the four sectors (household, commercial, transportation, and industrial) combine both efficiency projections and growth projections. When we aggregate these sector totals, we obtain the total for the whole economy. Factoring in the supply scenario for these sectors gives the results in Table 3-15.

The 126-Quad Demand Scenario

In the 126-quad demand scenario we assume a population in the year 2000 based on a fertility rate of 1.9 children per female. This divergence has very little effect on projections of labor force, hours worked, or the gross national product (GNP) in 2000. As a consequence, it is the technological assumptions that explain the major differences between the energy demands in the two scenarios.

The derivation of demand estimates for the four basic sectors, household, commercial, transportation, and industrial, are given beginning on page 81.

Table 3-14.

Total gross industrial energy demand
(quads)

Sector	1974	1985	2000
Manufacturing	18.5	23.4	31.1
Feedstocks and captive coke	4.9	6.0	8.5
Mining	2.3	2.9	3.1
Construction	1.2	1.5	1.7
Total	26.9	33.8	44.4

Table 3-15.
Summary of total and sector energy inputs, 1975 and 2000 (101 quads)

Sector	Total	Coal	Oil	Gas	Electricity	Geothermal, solar, etc.
1975						
Transportation	18.6		17.9	0.6	0.1	
Residential-commercial	25.8	0.2	5.7	7.6	12.2	
Industrial	25.9	3.8	5.5	8.5	8.0	
Total	70.3	4.0	29.1	16.7	20.3	
2000						
Transportation	22.2	0	21.2	0.6	0.4	0
Residential-commercial	34.4	0.3	0.9	6.0	27.2	0
Industrial	44.4	10.8	3.4	9.3	19.0	2.0[a]
Total	101.0	11.1	25.5	15.9	46.6	2.0

[a]Includes solar and geothermal for all uses—household, commercial, and industrial.

Household Energy Demand

Our calculations of household energy demand for the 126-quad scenario
follow the same procedure that we detailed in the 101-quad scenario, but it
diverges on certain key assumptions:

1. In the 101-quad scenario, we assumed a retrofit rate of 2 percent on old
houses; here we assume a 1 percent rate. This implies 14.4×10^6 retrofits in
2000. Total demand is then

insulation only for 90 percent of the homes:

$(0.9) (14.4 \times 10^6) (0.80) (118.1 \times 10^6) = 1.2$ quads

insulation and heat pump for 10 percent of the homes:

$(0.1) (14.4 \times 10^6) (0.80) (0.65) (118.1 \times 10^6) = 0.1$ quad

where 0.80 is the energy efficiency, 118.1×10^6 Btu/unit is the energy intensity, and 0.65 is the energy efficiency improvement for the heat pump.

2. This retrofit rate implies that 29×10^6 old houses are not retrofitted. For
these houses total demand is $(29 \times 10^6) (118.1 \times 10^6) = 3.4$ quads.

3. In the 101-quad scenario we assumed that heat pump installation in new
houses (we assume 58.6×10^6 built after 1975) reaches 25 percent of total;
here we assume a 10 percent installation of heat pumps. This results in a total
new house demand of 5.4 quads, as follows:

(no heat pump) $(0.9) (58.6 \times 10^6) (0.80) (118.1 \times 10^6) = 5.0$ quads

(with heat pump) $(0.1) (58.6 \times 10^6) (0.80) (0.65) (118.1 \times 10^6) = 0.4$ quad

4. Total demand for heating and cooling is thus 10.1 quads. But we increase
energy intensity by 1 percent annually to allow for increased market penetration of air-conditioning in both new and old units. This results in an energy
intensity at about the 1975 level and a total energy demand of 13.0 quads.

5. Finally, we assume a substantial increase in appliances, at 2 percent
annually. This gives a household demand total in 2000 of 16.1 quads, as compared with 12.4 quads for the 101-quad scenario.

The household energy demand projections are summarized in Table 3-3.

Commercial Energy Demand

The 126-quad commercial scenario diverges from the 101-quad scenario in
the following assumptions:

1. We assume a 1 percent retrofit rate instead of 2 percent. This results in

3×10^9 square feet of retrofitted commercial space. Total demand for retrofits is then

(no heat pump) (0.9) (3.0×10^9) (0.90) $(0.21 \times 10^6) = 0.5$ quad
(with heat pump) (0.1) (3.0×10^9) (0.90) (0.65) $(0.21 \times 10^6) = 0.04$ quad

where 0.90 is the energy efficiency, 0.21×10^6 Btu/square foot is the energy
intensity, and 0.65 is the energy efficiency improvement for the heat pump.
Nonretrofitted space totals 12.2×10^9 square feet. At the 1975 energy intensity of 0.21×10^6 Btu/square foot, this gives an energy demand of 2.6
quads. The energy demand for old commercial space is 3.1 quads.

2. For new commercial space we assume that the heat pump penetrates to
only 10 percent of the market. This results in a total demand of 5.3 quads:

(no heat pump) (0.9) (29.4×10^6) (0.90) (0.21×10^6) $= 5.0$ quads
(with heat pump) (0.1) (29.4×10^6) (0.90) (0.65) $(0.21 \times 10^6) = 0.3$ quad

$$\overline{ 5.3 \text{ quads}}$$

where the projected new commercial footage is 29.4×10^6 square feet.

3. For appliances we increase energy intensity 0.5 percent annually to allow
for increased market penetration. This results in a total appliance demand of
6.2 quads.

4. Finally, we drop the assumption of a shift in the composition of commercial energy demand (a shift away from schools to medical care and
greater expenditures for commercial amusements).

The commercial energy demand projections are summarized in Table 3-5.

Transportation Sector
In the 101-quad scenario, the assumption was made that automobile usage
had reached a point of saturation and would decline. The 126-quad assumption is quite different. The automobile stock increases from 0.67 car per person over the age of 16 to 0.71 in 1985 and 0.77 in 2000. In addition, the
annual mileage driven per car increases from 10,000 in 1975 to 11,000 in
1985 to 12,000 in 2000. (The 101-quad assumption was that the annual
mileage per car stayed at 10,000.) No change was made in fuel efficiency
factors from the 101-quad case; total fuel consumption in 2000 becomes 8.8
quads.

We assumed here that air transport grows at a rate 2 percent a year faster
than the GNP. This yields a total energy demand of 4.0 quads in 1985 and
7.7 quads in 2000 for the air transport sector.

For the remaining three categories of transport, (i) service trucks, (ii) truck and bus and rail freight, and (iii) ship and barge and pipeline transport, it is assumed that growth approximates the GNP growth (3.8 percent to 1985, 2.8 percent thereafter to 2000). This assumption is identical to that in the 101-quad scenario, but lower efficiency factors are used.

The transportation energy demand projections are summarized in Table 3-16.

Industrial Demand

In the 101-quad scenario we drew upon a detailed IEA study of the industrial sector. For the 126-quad scenario we have not tried to duplicate this level of analytical detail. Instead we have followed a more generalized analysis.

We assume that industrial growth approximates the GNP path, at 3.8 percent annually, until 1985, and that it then runs ahead of the GNP until 2000. Efficiency factors are assumed to average 10 percent by 1985 and 20 percent by 2000. Total demand increases from 25.9 quads in 1975 to 33.8 quads in 1985 to 52.5 quads in 2000.

Industrial uses of energy are divided into three broad categories: (i) process heat, (ii) electric drive, and (iii) lighting and feedstocks.

The most energy-intensive use of process heat is in the iron and steel and aluminum industries. In these industries we assume an efficiency improvement of 10 percent by 1985 and 20 percent by 2000. In the iron and steel industries we project a decrease in energy demand due to lighter-weight automobiles. Our estimate is a 2.2 percent growth rate to 1985 and a 2.0 percent rate thereafter. This yields a total energy demand for iron and steel of 2.6 quads by 1985 and 3.1 quads by 2000. In the aluminum industry we project continued rapid new product uses and an annual growth rate of 5 percent. The aluminum energy demand totals 1.3 quads by 1985 and 2.1 quads by 2000.

The less energy-intensive uses of process heat show perhaps the greatest potential for energy saving in the industrial sector. We are assuming that the energy demand per unit is reduced by 15 percent by 1985 and 35 percent by 2000. Combining these efficiencies with the assumption that growth in these industries will approximate the GNP path until 1985 and then run ahead of the GNP until 2000, we find that the total demand becomes 13.8 quads in 1985 and 18.5 quads in 2000.

Table 3-16.
Transportation energy demand projections: 126-quad scenario

Parameter		1975	1985	2000
Totals—energy inputs	$(\times 10^{15}$ Btu)	18.6	21.7	28.1[b]
Automobiles	$(\times 10^{6}$ vehicles)	105	126	152
	$(\times 10^{12}$ vehicle-miles)	1.05	1.39	1.82
	(Miles per gallon)	14	20	27
	$(\times 10^{15}$ Btu)	9.8	9.1	8.8
Service trucks	$(\times 10^{9}$ vehicle-miles) (%/year)[a]	99 (3.8)	144 (2.8)	217
	(Miles per gallon)	11	14	18
	$(\times 10^{15}$ Btu)	1.3	1.5	1.7
Truck and bus and rail freight	$(\times 10^{9}$ ton-miles) (%/year)[a]	503 (3.8)	731 (2.8)	1106
	(Efficiency factor)	1.0	1.0	0.9
	$(\times 10^{3}$ Btu/ton-mile)	7.1	6.8	6.3
	$(\times 10^{15}$ Btu)	3.6	5.0	7.0
Air transport	(Efficiency factor)	1.0	1.0	0.9
	$(\times 10^{15}$ Btu) (%/year growth)[a]	2.4 (5.8)	4.0 (4.8)	7.5
Ship and barge and pipeline transport	(Efficiency factor)	1.0	1.0	0.9
	$(\times 10^{15}$ Btu) (%/year growth)[a]	1.5 (3.8)	2.1 (2.8)	3.0

[a]Growth factors (in percent per year) are shown in parentheses for service and freight sectors for each time period.
[b]Totals may not add as a result of rounding.

Electric drive and lighting are projected to increase rather rapidly as a result of increased automation and industrial lighting. We project a 4.8 percent increase to 1985 and a 3.8 percent annual increase from 1985 to 2000. Combining these growth rates with efficiency improvements of 10 percent and 15 percent gives total energy demands of 9.9 quads in 1985 and 19.0 quads in 2000.

We estimate that feedstocks will grow at about the GNP growth rate. Assuming efficiency rates of 5 percent and 15 percent yields totals of 6.2 quads in 1985 and 9.8 quads in 2000 for feedstocks (see Table 3-17).

Total Energy Demand

When we aggregate the sectoral totals and factor in the supply scenario, we generate the data in Table 3-18.

Table 3-17.
Industrial energy demand projections: 126-quad scenario

126-quad scenario		1975	1985	2000
Total	$(\times 10^{15}$ Btu$)^a$	25.9^b (3.8)	33.8 (3.8)	52.5
	(Average efficiency factor)	1.0	0.9	0.8
Iron and steel	$(\times 10^{15}$ Btu$)^a$	2.3 (2.2)	2.6 (2.0)	3.1
	(Efficiency factor)	1.0	0.9	0.8
Aluminum	$(\times 10^{15}$ Btu$)^a$	0.9 (5.0)	1.3 (4.0)	2.1
	(Efficiency factor)	1.0	0.9	0.8
Other process heat	$(\times 10^{15}$ Btu$)^a$	11.2 (3.8)	13.8 (3.8)	18.5
	(Efficiency factor)	1.0	0.85	0.65
Electric drive and lighting	$(\times 10^{15}$ Btu$)^a$	6.9 (4.8)	9.9 (4.8)	19.0
	(Efficiency factor)	1.0	0.9	0.85
Feedstocks	$(\times 10^{15}$ Btu$)^a$	4.5 (3.8)	6.2 (3.8)	9.8
	(Efficiency factor)	1.0	0.95	0.85

[a]Growth rates (in percent per year) are shown in parentheses for each sector and time interval.
[b]Totals may not add as a result of rounding.

Table 3-18.
Summary of total and sector energy inputs, 1975 and 2000 (126 quads)

Sector	Total	Coal	Oil	Gas	Electricity	Geothermal, solar, etc.
1975						
Transportation	18.6	0	17.9	0.6	0.1	
Residential-commercial	25.8	0.2	5.7	7.6	12.2	
Industrial	25.9	3.8	5.5	8.5	8.0	
Total	70.3	4.0	29.1	16.7	20.3	
2000						
Transportation	28.1	0	27.1	0.6	0.4	
Residential-commercial	45.3	0.3	1.3	6.2	37.5	
Industrial	52.5	11.2	6.4	9.1	24.0	2.0[a]
Total	125.9	11.5	34.8	15.9	61.9	2.0

[a] Includes geothermal and solar for all uses—household, commercial, and industrial.

REFERENCES

1. Bureau of the Census, *1972 Census of Manufacturers,* "Fuels and Electric
   Energy Consumed," Special Report Series MC-72, SR-6, U.S. Government
   Printing Office, Washington, D.C., 1973.

2. See Table 1-9.

3. The impact of higher fuel prices has been to greatly accelerate the installation of insulation. The Department of Commerce reported that 3×10^6
   units were retrofitted in the first half of 1977, compared to 750,000 in the
   same period of 1976 [*Washington Post,* November 5, 1977, p. E17]. This
   highly accelerated rate is expected to taper off within a few years. Costs
   per house ran about $540 to increase insulation from none to R-33 (over
   10 inches), and the pay-out periods ran from 1.3 to 2.4 years, depending
   on the type of fuel used. These are Owens-Corning Fiberglas Corporation
   estimates, quoted in the same article. For houses with partial insulation,
   the pay-out period would be longer, since the annual energy savings would
   be less. For our purposes in estimating household fuel demand in 2000,
   the annual rate of retrofit is not as important as the share of the housing
   stock that is adequately insulated.

4. Center for Advanced Computation, "Energy Intensive Ranked Sector
   Table," University of Illinois, Urbana, Document No. 128 (Technical
   Memorandum 23), 1971.

5. E. L. Allen *et al., U.S. Energy and Economic Growth, 1975-2010,* Publication ORAU/IEA-76-7, Institute for Energy Analysis, Oak Ridge Associated Universities, Oak Ridge, Tennessee, 1976.

6. Bureau of the Census, *Annual Survey of Manufacturers, 1974,* "Fuels and
   Electric Energy Consumed," Series M74(AS)-4.2, U.S. Government Printing Office, Washington, D.C., September 1976.

4

**ENERGY SUPPLY
SCENARIOS**

The 101-Quad Supply Scenario

Under the 101-quad supply scenario for the year 2000, the major supply shift is estimated to be toward electricity and away from natural gas and oil, which will be scarcer in relation to demand. About 80 percent of the process heat generated in the industrial sector is now supplied by the burning of oil and natural gas, and much of this can be supplanted by coal. Since the proportion of the total gross energy expected to be utilized in the generation of electricity increases to over 46 percent by the year 2000, a breakdown into energy sources used for the generation of electricity and for other purposes is shown in Table 4-1.

We do not anticipate any other significant shifts in the national fuel pattern between now and 2000. Supplies of natural gas and domestic oil are expected to be close to present levels, aided by the exploitation of Alaskan reserves. Solar energy, geothermal energy, biomass, wind, and other exotics are expected to contribute a combined total of perhaps 4 quads under current research, development, and demonstration (RD&D) emphasis (see Table 4-1). The supply sources and the relevant assumptions are discussed briefly below.

Hydro Power

The generation of electricity from hydro power is expected to expand only very moderately by 2000. Although the nation possesses significant unex-

Table 4-1.
Fuel inputs by source, 1975 (actual) and 2000 (estimated) for the 101-quad scenario (10^{15} Btu)

Generation of	Total	Hydro power	Solar and geothermal energy	Nuclear power	Coal	Natural gas	Liquids Sub-total	Shale and biomass	Domestic	Imported
1975										
Total	70.6[a]	3.2	0	1.8	12.8	19.9	32.6[a]	0	19.7	12.7
Electricity	20.2	3.2	0	1.8	8.8	3.2	3.2	0	3.1	0.1
Direct fuels	49.8	0	0	0	4.0	16.7	29.1	0	16.6	12.6
2000										
Total	101.0	4.1	3.5	14.5	31.6	18.1	29.2	2.5	19.0	7.7
Electricity	46.6	4.1	1.5	14.5	20.6	2.2	3.7	0	0.9	2.8
Direct fuels	54.5	0	2.0[b]	0	11.1	15.9	25.5	2.5	18.1	4.9

[a]Includes 0.3 quad miscellaneous or unaccounted for.
[b]Direct thermal heating.

ploited hydro potential, the majority of attractive sites are in national parks or other scenic areas and therefore are unlikely to be used. We do not anticipate a move to dam such places as the Grand Canyon. Some expansion, particularly of pumped storage, is expected, but the net new contribution by 2000 is not forecast to exceed 1 quad above current levels. A concerted drive to reactivate small units on New England rivers could raise this estimate by a fraction of a quad.

Solar and Geothermal Energy

Under present research and development (R&D) emphasis, we believe there will be a small but significant contribution from solar and geothermal energy resources by 2000. Geothermal resources capable of exploitation with current technologies are largely in the West; prime solar areas are in the Southwest.

The contribution of solar energy up to 2000 is expected to be primarily for the supplemental heating of homes and commercial establishments and for heating water. President Carter's energy plan of April 20, 1977, projects the use of solar energy in 2.5×10^6 homes by 1985. If solar energy is to supply 1 quad of energy, about 8×10^6 homes would need to be equipped with solar panels and heat storage systems. In 1975 dollars and at present solar equipment prices, this would require the investment of about $40 billion (1). Future technical advances in solar energy processes could cut these costs significantly, and R&D is now being conducted to this end; at the same time, generating costs by conventional means (coal and nuclear power) are likely to increase. To penetrate the market significantly, however, solar units will probably require long-term storage capacity and this has yet to be developed.

Apart from what could be characterized as "active" solar energy systems (collectors, fans, pumps, and other mechanical devices), there is obvious potential for energy saving through "passive" solar installations. Passive solar energy is illustrated by buildings designed so that the structure serves as an efficient solar collector and short-term storage unit. In such buildings, thick walls, natural ventilation systems, and evaporating cooling systems provide warm-weather comfort. Where it is possible to install such systems, they will undoubtedly be increasingly used.

Geothermal resources can be used to provide energy if the heat contained in geological formations is extracted; this heat can then generate electric power or provide industrial process, space, and hot-water heat. Most known hydrothermal resources are concentrated in a few western states, particularly in California but also in New Mexico and Idaho. Anticipated electric power

generation from these resources a decade or so hence continues to be scaled
down in California; estimates have fallen from 7000 megawatts (electric)
for 1990 to 2000 megawatts (electric) for 1985 (2). There are a number of
serious environmental problems associated with geothermal exploitation, and
more rapid development awaits their solution. The recovery of extensive
methane and geothermal reserves from the Gulf of Mexico is primarily a technical problem.

Nuclear Power

Projections by IEA of the nuclear power contribution by 2000 are based very
largely on a regional compilation of expected new plants compiled by the
Bureau of Mines (3). The energies range from 14.5 quads in our 101-quad
scenario to 22.6 quads in the 126-quad scenario. In President Carter's National Energy Plan (NEP) an increase to 7.6 quads in 1985 was estimated
from 2.0 quads in 1976 (4). We made some modifications in the Bureau of
Mines data in arriving at the IEA estimates for 2000, since regional planning
in relation to estimated demand is uneven. Hence, some nuclear plants were
dropped out of the planning schedule in cases where electric power expansion plans appeared to be running ahead of regional demands.

There is much uncertainty associated with estimates of future nuclear capacity. Clearly, the electric utility industry's forward planning gives a greatly
enhanced role to nuclear power. The Carter Administration acknowledges the
need for rapid expansion of nuclear power in the NEP. Yet vigorous opposi-
tion by environmental groups is hamstringing plant construction and acceler-
ating construction costs.

Although nuclear power plants are still being built, the current lack of
orders for future delivery is a matter of concern. It is possible that the enact-
ment of proposed Carter Administration legislation will reduce construction
times in spite of environmental delays. Also, the recent unanimous decision
by the Supreme Court, warning federal judges not to impose regulatory obstacles to the construction and operation of two nuclear plants, may help (5).

Coal

We anticipate a very sharp increase in coal's contribution to energy supplies
between now and 2000 (see also Appendix B). In 1975, about 12.8 quads
were consumed, or 18.1 percent of the total primary energy. In the 101-quad
scenario, we project 31.6 quads to be consumed by 2000, which would represent 31.3 percent of the total primary energy. Depending on the ratio of

western coal (which has a lower heat content than eastern coal), this could represent about 825 x 10^6 to 850 x 10^6 tons by 1985 and 1.4 x 10^9 tons in 2000. The 1985 projection is well within that implied in the NEP. Under the 101-quad scenario, we assume that environmental and transportation problems associated with boosting coal output will keep its growth well below the levels contemplated in the NEP. A major stimulus to increasing coal output had been expected to be its price attractiveness compared to that of other primary energy sources. As a result of the late 1973 oil embargo of the Organization of Petroleum Exporting Countries (OPEC) and the coal strike of 1977-1978, the prices of oil and natural gas escalated much more rapidly than coal prices. By early 1976, coal was cheaper on a Btu basis than either of its principal competitors (see Figure 4-1). However, it is not certain that this cost edge will continue, in view of the highly inflationary wage settlement reached with the United Mine Workers early in 1978, which in itself will raise the price of coal by $4 a ton or more in eastern and midwestern markets.

In addition, the 1977 amendments to the Clean Air Act (6) require the use of the best available control technology (BACT) on stack emissions so as to reduce their sulfur content. This amendment will require the installation of stack scrubbers on all new power plants, which will increase the investment

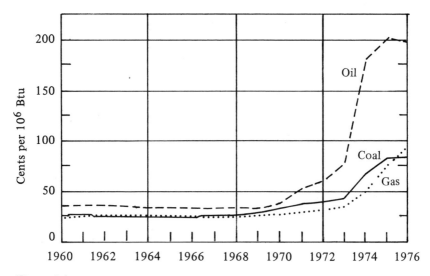

Figure 4-1.
Trends in the prices of fossil fuels sold to steam electric plants (cents per 10^6 Btu, 1960-1976)

required to construct a coal-fired utility plant some 20 to 25 percent. Other new federal regulations bring with them increasing costs for the restoration of strip-mined coal fields, delays in the leasing of federally owned coal fields, and perhaps stricter emission standards. Hence, there is much greater uncertainty about coal's future today than there was early in 1977. Finally, controlled prices for competitive fuels may continue to adversely affect the demand for coal, at least for another 5 years.

Natural Gas

We anticipate only a slight decrease in the availability of domestic natural gas during the remainder of this century. Higher prices have already begun to stimulate domestic drilling. Alaskan supplies will almost certainly be delivered to the lower 48 states in the 1980s, completion of a pipeline from Mexico's Reforma field is expected within a few years, and liquefied natural gas could be coming in not only from Algeria but from other offshore producers as well. Gas is a premium fuel in the sense that it provides heat with high efficiency, convenience, and cleanliness. There are additional potential natural gas supplies such as those in geothermal formations in the Gulf of Mexico and in Devonian shales, which would become commercially available in the future, given successful RD&D. The major uncertainty is probably the question of the continuation of price controls on interstate shipments and the prospect of new federal regulations on the price of intrastate sales.

Petroleum

Under the 101-quad scenario, we are making two key assumptions on domestic petroleum supplies in the future: (i) that domestic production of crude, aided by enhanced oil recovery (EOR) and by new supplies from Alaska and offshore, will hold output close to current levels, and (ii) that shale oil and biomass will contribute about 2.5 quads of liquids in 2000. On the demand side, we assume that mandated miles-per-gallon requirements for private automobiles and light trucks, greatly improved passenger mile performance by commercial aircraft, and the mandated shifts from oil to coal contained in President Carter's NEP proposal will effectively restrict demand to slightly less than 30 quads. The demand impact of growing scarcity and higher prices for petroleum would increasingly restrict its consumption to the transportation sector. We make no allowance in this scenario for liquids from coal. The technology to produce synthetics from coal has been available since

World War II, but at prices for petroleum anticipated between now and the
year 2000 no appreciable production of synthetics from coal is likely.

Overall

Energy supplies to meet the 101-quad scenario in 2000 would require only
modest inputs from new technologies. Solar power and geothermal energy
would supply perhaps 3.5 quads. If a liberal allowance is made for oil shale
and biomass, the combined contribution of these three sources would be 6
percent of the total demand, or about 6 quads. Given a reasonably effective
conservation effort, by 2000 primary reliance for energy supplies would continue to be on coal, oil, natural gas, nuclear power, and hydro power—perhaps
95 percent of the energy supplies in that year (see Table 4-2).

The 126-Quad Supply Scenario

In increasing the supply scenario from 101 to 126 quads in the base case,
there are two major changes: (i) much greater consumption (some 5.8 quads)
of petroleum in the transportation sector, and (ii) the need to provide 15.8
quads more of electric power. The greater use of energy in the transportation
sector is the result of increased estimates of the number of automobiles likely
to be on the road as well as of the increased average mileage driven per car
over the case for the 101-quad scenario. There are no technical fuel substitution problems in this adjustment, because with unimportant exceptions

Table 4-2.

Estimated major energy inputs in 2000
(101-quad scenario)

	Percent	
Fuel	Of total	Cumulative
Coal	31.7	31.3
Domestic and imported petroleum	29.2	60.2
Natural gas	18.1	78.1
Nuclear power	14.5	92.5
Hydro power	4.1	96.6
All other (biomass, solar power, wind, geothermal energy)	6.0	101.0

petroleum must be used to power transport vehicles (land, sea, and air). Since domestic oil production is limited, the additional supplies would need to be imported, and this could create additional balance of payments problems for the United States. At early 1978 landed prices for petroleum of about $14.50 a barrel, the added 5.8 quads would cost over $15 billion (in 1978 prices). Moreover, this level of U.S. import increase could adversely affect petroleum supplies for other western industrialized nations, thereby raising prices and causing economic disruption. The provision of additional electric power by 2000 would not appear to be supply-limited unless new governmental regulations and actions by environmental groups succeed in holding electric power output below demand growth.

The lion's share of base-load electric utility plants could be either coal-fired or nuclear-powered, and there will be a contest between these two sources over the next 25 years. Nuclear plants seem to enjoy a modest cost advantage at present. At the moment, there appears to be a *de facto* moratorium on placing new orders for nuclear plant equipment. There are also serious environmental concerns that could inhibit expansion of coal-fired plants until the associated pollution problems are brought under control.

Current Federal Power Commission data indicate a vigorous expansion of nuclear-powered electricity-generating capacity (7). Given the uncertainties, we have divided the requirement for added electric power in the 126-quad scenario between nuclear power and coal roughly on a 50-50 basis to reflect the probable toss-up fuel type decisions to 2000 (see Table 4-3).

Table 4-3.
Fuel input by source, 1975 (actual) and 2000 (estimated) for the 126-quad scenario (10^{15} Btu)

Generation of	Total	Hydro power	Solar and geothermal energy	Nuclear power	Coal[a]	Natural gas	Liquids			
							Sub-total	Shale and biomass	Domestic	Imported
1975										
Total	70.6[b]	3.2	0	1.8	12.8	19.9	32.6[b]	0	19.7	12.7
Electricity	20.2	3.2	0	1.8	8.8	3.2	3.2	0	3.1	0.1
Direct fuels	49.8	0	0	0	4.0	16.7	29.1	0	16.6	12.6
2000										
Total	126.0	4.1	3.5	23.6	39.3	18.1	38.4	2.5	19.0	16.9
Electricity	62.4	4.1	1.5	23.6	27.8	2.2	3.6	0	0.9	2.8
Direct fuels	63.6	0	2.0[c]	0	11.5	15.9	34.8	2.5	18.1	14.1

[a]Coal consumption, depending upon the proportion of low-heat western coal and lignite consumed, would run between 1.7×10^6 and 1.8×10^9 tons in 2000.

[b]Includes 0.3 quad miscellaneous or unaccounted for.

[c]Direct thermal heating.

REFERENCES

1. A. M. Bueche, *The Hard Truth About Our Energy Future,* General Electric Company, New York, March 1977. An optimistic appraisal of solar energy's future role is given in the Office of Technology Assessment, Congress of the U.S., *Application of Solar Energy to Today's Energy Needs,* Washington, D.C., 1977.

2. A high estimate is given in California State Resources Agency, *Energy Dilemma,* Sacramento, June 1973. The more recent estimate is from Spurgeon M. Keeney *et al., Nuclear Power Issues and Choices,* Ballinger, Cambridge, Massachusetts, 1977, p. 141.

3. John S. Corsentino, *Projects to Expand Fuel Sources in Western States,* Bureau of Mines, Publication 1C8719, U.S. Government Printing Office, Washington, D.C., 1976; Bureau of Mines, Eastern Field Operations Center, *Projects to Expand Fuel Sources in Eastern States,* Publication 1C8725, U.S. Government Printing Office, Washington, D.C., 1976.

4. Executive Office of the President, *The National Energy Plan,* Washington, D.C., 1977, p. 96.

5. See *New York Times,* April 4, 1978, p. 1.

6. *The Clean Air Act as Amended August 1977,* Serial No. 95-11, 95th Congress, 1st Session, U.S. Government Printing Office, Washington, D.C., 1977.

7. Federal Power Commission, News Release No. 22763, National Electric Reliability Council, "Fossil and Nuclear Fuel for Electric Utility Generation: Requirements and Constraints, 1976-85," Washington, D.C., 1976, table 4, p. 13.

5

**REGIONAL
ENERGY DEMANDS**

Introduction

In this chapter we estimate the regional energy demand and supply patterns for the year 2000. We consider each of four demand sectors—household, commercial, transportation, and industrial. We also estimate electric power supply and demand in each region, based on a detailed analysis of the expansion plans of the public utilities. The sector considered in greatest detail is industrial demand, since this is the largest single energy-consuming sector and the factor least studied in other regional surveys (1).

The pattern of U.S. energy consumption per capita varies widely from one Census Bureau region to another, with the highest use in the West South Central region and the lowest in the New England and Mountain regions. The geographical variation in per capita energy use in the household, commercial, and personal transportation sectors can largely be explained in terms of variations in climate and population density. The largest regional variation in per capita energy use, however, is due to differences in the industrial energy sector. These differences result from the large concentration of energy-intensive manufacturing in a few locations, as shown below in the section on "Industrial Demand."

In developing regional projections, we have broken down only the 101-quad scenario. Corresponding regional projections based on the 126-quad total could be calculated, but they have been omitted here because of the laborious nature of the process.

Total Regional Energy

We begin with an analysis of the gross regional energy consumption in 1975. The latest available Bureau of Mines (BOM) comprehensive regional breakdown is for 1973 (2), but preliminary 1975 regional data on nuclear and hydro power have been published by the Federal Power Commission (3). Natural gas, petroleum, and coal data are given in BOM industry surveys on these commodities. The results are shown in Table 5-1.

Using the 1975 regional population data and the energy consumption estimates, we calculate a per capita energy intensity (in 10^9 Btu). The regional differences are assumed to persist to the year 2000, because they are dependent on permanent or long-term factors such as climate. Our estimates of industrial regional shifts to 2000 show only small changes. To calculate a regional per capita intensity factor for 2000, we project the national per capita intensity based on IEA control totals of 101 quads and 245.1×10^6 people. This translates into a 23 percent increase over the 1975 national energy intensity, a consumption per capita of 0.334×10^9 Btu. Assuming an equiproportional increase in each region, we can estimate the regional intensities. Multiplying these intensities by the regional population projections generates the regional energy demand totals.

Households

Government-compiled data (BOM) on regional household energy use are available for 1973. However, the BOM reports regional and state energy demand for combined household and commercial uses. Separate totals for household and commercial uses must be determined. Also, the BOM sector consumption data for electricity are for net energy use, not gross. To compute the gross energy use, the total energy used for electricity generation is distributed with respect to the consuming sector. Adding electricity-generating uses to nonelectricity energy uses in the household and commercial sectors yields a total of 25.9 quads for 1973. To separate the household and commercial sectors, we use IEA estimates (4), which are 63 percent for households or 16.3 quads and 37 percent for commercial or 9.6 quads.

Our basic strategy is to extrapolate regional intensities to the year 2000, modified to reflect household conservation factors calculated in Chapter 3. The overall household energy intensity for the 72×10^6 homes in 1973 was 237×10^6 Btu/unit; the projected intensity for 2000 is 214×10^6 Btu/unit.

Table 5.1.
Regional gross energy consumption[a]

Region[b]	1975		2000		
	Consumption (quads)	Per capita energy intensity (x 10⁹ Btu)	Population (x 10⁶)	Per capita energy intensity (x 10⁹ Btu)	Adjusted consumption (quads)
U. S. total	71.1	0.334	245.1	0.412	101.0
New England	3.1	0.254	13.7	0.314	4.3
Middle Atlantic	10.4	0.279	40.4	0.344	13.9
East North Central	14.0	0.341	45.3	0.419	19.0
West North Central	5.4	0.323	18.6	0.398	7.4
South Atlantic	9.4	0.279	41.1	0.343	14.1
East South Central	4.7	0.348	15.7	0.427	6.7
West South Central	12.1	0.579	24.8	0.714	17.7
Mountain	3.7	0.385	12.5	0.472	5.9
Pacific (including Hawaii and Alaska)	8.3	0.289	33.0	0.355	11.7

[a] As a result of rounding, figures may not add to totals.

[b] The regional grouping of states (Census Bureau breakdown) is given below. New England: Maine, New Hampshire, Vermont, Massachusetts, Rhode Island, and Connecticut; Middle Atlantic: New York, New Jersey, and Pennsylvania; East North Central: Ohio, Indiana, Illinois, Michigan, and Wisconsin; West North Central: Minnesota, Iowa, Missouri, North Dakota, South Dakota, Nebraska, and Kansas; South Atlantic: Delaware, Maryland, District of Columbia, Virginia, West Virginia, North Carolina, South Carolina, Georgia, and Florida; East South Central: Kentucky, Tennessee, Alabama, and Mississippi; West South Central: Arkansas, Louisiana, Oklahoma, and Texas; Mountain: Montana, Idaho, Wyoming, Colorado, New Mexico, Arizona, Utah, and Nevada; and Pacific: Washington, Oregon, California, Hawaii, and Alaska.

This implies a conservation coefficient of 214 x 10^6/237 x 10^6 = 0.9. This result reflects two offsetting trends: (i) the conservation effort that is, for the most part, a function of rising energy prices and (ii) the proliferation of new energy-intensive appliances.

This conservation coefficient is applied interregionally. We also assume that the present regional intensity differences will not change, because they are largely a function of persistent factors. The household energy intensity for 2000 is thus a function of

$$0.9 \left(\frac{E_{1973}}{H_{1973}}\right)$$

where H is the number of households and E is the energy consumed (in quads).

To project the number of households in 2000, we begin with the IEA regional population projections and develop assumptions about household size and composition. The IEA projects a secular reduction in household size to 2.36 persons per household in 2000 (5). Not every person is a member of a household; there are persons in institutions, such as hospitals, homes for the aged, army barracks, and prisons. The nonhousehold population comprises about 2 percent of the total (6). The household total for 2000 is given by

$$\frac{0.98 \times P_{2000}}{2.36}$$

where P is the population.

Total household energy consumption for 2000 is a function of the household total and the household intensity:

$$E_{2000} = 0.9 \left(\frac{E_{1973}}{H_{1973}}\right)\left(\frac{0.98 \times P_{2000}}{2.36}\right)$$

Regional projections are calculated in Table 5-2. In the case for New England, with a 1973 household energy intensity of 0.308 x 10^9 Btu and a population in 2000 projected at 13.7 x 10^6, consumption for 2000 is

$$E_{2000} = 0.9 \left(\frac{1.2}{3.9 \times 10^6}\right)\left(\frac{0.98 \times 13.7 \times 10^6}{2.36}\right) = 1.6 \text{ quads}$$

Table 5-2.
Regional household energy consumption, 1973 and 2000[a]

Region	Households and energy use, 1973			Households and energy use, 2000	
	Numbers (x 10⁶)	Energy consumption (quads)	Energy intensity (x 10⁹ Btu)	Numbers (x 10⁶)	Energy consumption (quads)
U.S. total	68.7	16.3	0.237	102.0	21.8
New England	3.9	1.2	0.308	5.7	1.6
Middle Atlantic	12.5	3.3	0.264	16.8	4.0
East North Central	13.2	3.7	0.280	18.8	4.7
West North Central	5.5	1.5	0.273	7.7	1.9
South Atlantic	10.5	1.9	0.181	17.1	2.8
East South Central	4.2	0.8	0.190	6.5	1.1
West South Central	6.5	1.4	0.215	10.3	2.0
Mountain	2.9	0.8	0.276	5.2	1.3
Pacific	9.4	1.6	0.170	13.7	2.1

[a]As a result of rounding, figures may not add to totals.

Commercial

When the 1973 household and commercial total is separated, the commercial total is 9.6 quads. But, unlike the household case for which regional totals are available, there are no regional data on commercial footage. We use population data as a proxy, because the correlation between population and commercial space is likely to be close and regional differences are probably negligible.

Extrapolating regional intensities to the year 2000, we find that energy needs are modified to reflect the commercial conservation factors projected in Chapter 3. The 1973 commercial energy intensity is 0.046×10^9 Btu per person for the nation as a whole. This increases to 0.051×10^9 Btu per person by the year 2000. The increase is a function of two factors: (i) a combination of modest conservation measures plus a shift to less energy-intensive services resulting from an aging population and (ii) a 25 percent increase in commercial space per household. In combination, the commercial intensity increases 12 percent over that in 1973.

Regional commercial energy consumption for the year 2000 is a function of the regional population projections and the respective commercial intensities, that is,

$$E_{2000} = 1.12 \frac{E_{1973}}{P_{1973}} (P_{2000})$$

Regional commercial energy projections are calculated in Table 5-3. For example, for New England with a 1973 commercial energy intensity of 0.058×10^9 Btu per person and a population in 2000 projected at 13.7×10^6, consumption for 2000 is

$$E_{2000} = 1.12 \, (0.058 \times 10^9) \, (13.7 \times 10^6) = 0.9 \text{ quad}$$

Transportation

Transportation uses of energy include the personal (private and publicly owned) automobile; trucks (including service trucks) and buses; trains; ships, barges, and pipelines; and planes.

We begin our personal automobile fuel estimates for 2000 with a state-by-state survey of automobiles and gasoline consumption per vehicle in 1975

Table 5-3.
Regional commercial energy consumption, 1973 and 2000[a]

Region	Commercial energy use, 1973			Commercial energy use, 2000	
	Population (x 10^6)	Energy consumption (quads)	Energy intensity (x 10^9 Btu)	Population (x 10^6)	Energy consumption (quads)
U.S. total	209.8	9.6	0.046	245.1	12.6
New England	12.1	0.7	0.058	13.7	0.9
Middle Atlantic	37.4	1.9	0.051	40.4	2.4
East North Central	40.8	2.3	0.056	45.3	3.0
West North Central	16.6	0.8	0.048	18.6	1.0
South Atlantic	32.6	1.1	0.034	41.1	1.7
East South Central	13.3	0.4	0.030	15.7	0.5
West South Central	20.3	0.8	0.039	24.8	1.1
Mountain	9.2	0.4	0.043	12.5	0.7
Pacific	27.5	1.0	0.036	33.1	1.3

[a] As a result of rounding, figures may not add to totals.

based on Federal Highway Administration data (7). From these data, a regional average use of fuel per vehicle (intensity) is calculated. Using the Census Bureau Series III population projection to 2000, we estimate that 79.4
percent of the population will be 16 years of age or older and that 65 percent
of these will have automobiles in 2000. The impact of an aging population on
increased numbers of automobiles is shown in the following data:

	1975	2000
Total population (x 10^6)	213.6	245.1
Total 16 years or older (x 10^6)	155.6	194.7
Percent 16 years and over	72.8	79.4

Thus we estimate that there will be 127 x 10^6 cars by 2000, compared to
about 107 x 10^6 in 1975 (8). In our estimates we expect the miles-per-gallon
averages to improve from 14 in 1975 to 27 in 2000, and the number of miles
per car is held constant at 10,000 per year. The estimate of the miles-per-
gallon improvement is particularly important in arriving at the calculated
energy consumption by personal automobiles in 2000, since it falls from 9.8
quads in 1975 to 6.1 quads in 2000. The regional automobile energy breakdown in 2000 is shown in Table 5-4.

Truck and bus energy use estimated for 2000 also begins with a state-by-
state analysis of fuel consumption in 1975. Total numbers of trucks and
buses are added for each region, and an average fuel use per vehicle is calculated from Federal Highway Administration tables. The regional percentages
of national fuel use in 1975 are then adjusted to the year 2000, based on
projected shifts in regional populations. The calculated 2000 regional percentages are used to allocate the 7.45 quads of energy use derived from an
earlier IEA study [(4), pp. 89-91]. This calculation is an approximation, since
the amount of gasoline and diesel fuel used by service trucks and heavy-haul
trucks on highways is not separated from automobile use by the Federal
Highway Administration. Highway use accounted for about 97 percent of
the total consumption of motor fuels in 1975.

Energy use by railroads is projected to parallel the growth of rail freight in
ton-miles. The controlling percentage, the growth of the economy, is the
same expansion rate as used for trucks and buses. Data on the sales of distillate-
type and residual-type fuel oils for use by railroads have been compiled by
the BOM for 1975 on a state-by-state basis (9). The first adjustment that we
made was to adopt the 1975 regional percentages of fuel use to 2000 by ad-

Table 5-4.
Transportation regional energy consumption, 2000[a]
(quads)

Region	Totals	Autos	Trucks and buses	Trains	Ships, barges, and pipelines	Aviation fuel
Totals	22.2	6.10	7.45	1.15	2.90	4.60
New England	0.77	0.33	0.20	0.01	0.06	0.17
Middle Atlantic	2.64	0.90	0.52	0.06	0.41	0.75
East North Central	3.06	1.12	1.07	0.22	0.07	0.58
West North Central	1.78	0.46	0.88	0.14	0.03	0.27
South Atlantic	3.38	1.06	1.24	0.13	0.26	0.69
East South Central	1.56	0.42	0.65	0.08	0.29	0.12
West South Central	3.73	0.71	1.23	0.22	1.13	0.44
Mountain	1.42	0.32	0.65	0.16	0.0	0.29
Pacific	3.86	0.78	1.01	0.13	0.65	1.29

[a] As a result of rounding, figures may not add to totals.

justing for the anticipated shifts in population between 1975 and 2000. A
second adjustment made was for anticipated geographic shifts in coal production, since coal is the largest single commodity tonnage moved by railroads
(about 25 percent of the total). As a consequence, we adjusted the share of
the Mountain, West North Central, and East North Central regions upward.
The major adjustment was given to the Mountain region, since it was the only
one of the three that had both an above-average population increase and a
boost in the national coal share. However, these adjustments were not large,
since the total railroad energy use in 2000 is calculated to be 1.15 quads, up
from 0.54 quad in 1975.

We estimate that the energy consumption by ships, barges, and pipelines will
hold the same regional distribution pattern in 2000 that it did in 1975, since
waterborne traffic is unlikely to undergo significant redistribution. Growth
between 1975 and 2000, as in the case of trucks and buses and trains, is based
on general trends in economic expansion. Sales of distillate and residual fuels
for vessel bunkering were compiled by the BOM for 1975 on a state-by-state
basis (9). The largest single regional use is in the Gulf states, with the Pacific
Coastal states second. We have not attempted to adjust our 2000 control
total, 2.9 quads, for any regional differential that might follow from gasoline
sales to small maritime vessels, largely pleasure craft. This energy use is comparatively small and probably largely parallels the ship and barge consumption pattern by region.

Aviation fuel accounts for a significant share of the national energy consumption. The rate of growth of aviation fuel use between 1975 and 2000 is
expected to be the same as that of the industrial sector. In our calculation we
assume a modest efficiency improvement to 0.9 by 1985 and 0.8 by 2000.
This will raise the consumption of aviation fuels to 4.6 quads in 2000, compared to 2.4 quads in 1975. The regional distribution, however, is not expected to change significantly. An inspection of 1975 consumption shows
that most of the fuel is used in the coastal terminal centers—New York, California, Texas, and Florida; there is one additional major aviation fuel consumption center, namely, the hub at Chicago. Fuel consumption is more
closely linked to these transfer centers than to the regional distribution of
populations or manufacturing activity. The regional consumption breakdown
on aviation fuels is taken from the current estimates compiled by the Independent Petroleum Association of America (10). This estimate is based on
data compiled by the BOM and the Ethyl Corporation (see Table 5-4).

Industrial Demand

The location of industries, particularly those industries that use large quantities of energy, is the most important factor determining future regional energy requirements. Industry currently uses almost two-fifths of the total net energy, but this use varies widely from region to region. The West South Central region used nine times as much energy for each dollar of value added as New England and more than ten times as much industrial energy per capita. This energy use compares to a difference of less than 40 percent between the highest and lowest per capita use of nonindustrial energy (see Table 5-5).

Many factors determine where an industry will locate—the availability of raw materials, the markets to which they sell, the availability of labor, and other factors such as shipping routes, tax advantages, and the supply of cheap energy. The nearness of a cheap energy supply is unimportant to most industries, but it is important to those industries that use large amounts of energy. In general, the regions with the highest per capita industrial energy use are those where the price of energy is the lowest (Table 5-5).

We were not able to include many important variables in these estimates because we were unable to predict them. For example, with the exception of coal whose price is expected to be considerably lower in the West than in the East, we assumed no shift in the relative price of energy between regions. Also, we made little allowance for future regional differences in environmental standards, although these could have an important impact on the location of energy-intensive industries forced to convert to coal.

Our estimates show what the total energy requirements for industry would be for each region according to our projections of the growth rate, the energy/output ratios, and regional shifts within each industry. The results are shown in Table 5-6.

The regional distribution estimated for industry in 2000 appears quite similar to that of 1974. In general, the shift is expected to be away from the eastern to the western regions, although we believe that the share of the West South Central region will decline slightly. These estimates reflect anticipated regional labor and energy availabilities which account for some of the small differences in regional shares. Based on our current knowledge of coal and nuclear energy developments by region and on environmental laws (both of which adversely affect New England and the Middle Atlantic region), we anticipate no radical shifts in industrial activity between 1975 and 2000.

Table 5-5.
Regional industrial energy consumption and the price of energy, 1973

Region	Gross industrial energy per dollar value added (x 10³ Btu)	Per capita		Cost per 10⁶ Btu (cents)
		Net industrial energy (x 10⁶ Btu)	Nonindustrial energy (x 10⁶ Btu)	
U.S. total	64.1	97	117	48.4
New England	25.2	29	182	70.5
Middle Atlantic	45.0	62	166	63.7
East North Central	51.2	104	181	45.6
West North Central	47.1	65.4	203	37.7
South Atlantic	52.5	65.2	155	52.1
East South Central	82.6	106	167	39.9
West South Central	225.9	314	200	31.6
Mountain	112.2	95	213	34.9
Pacific	49.1	60	166	65.6

Table 5-6.
Regional industrial gross energy use, 1974 and 2000
(quads)

Region	Manufacturing[a]	Mining	Construction	Total	Total percent
1974					
New England	0.569	0.005	0.068	0.642	2.5
Middle Atlantic	3.236	0.068	0.217	3.521	13.7
East North Central	5.519	0.119	0.229	5.867	22.8
West North Central	1.040	0.161	0.096	1.297	5.0
South Atlantic	2.317	0.149	0.183	2.649	10.3
East South Central	1.865	0.086	0.073	2.024	7.9
West South Central	5.129	1.150	0.113	6.392	24.9
Mountain	0.598	0.326	0.050	0.974	3.8
Pacific	1.889	0.279	0.163	2.331	9.1
U.S. total	22.162	2.343	1.192	25.697	100.0

[a]Includes feedstocks and captive coke

Table 5-6.
Regional industrial gross energy use, 1974 and 2000
(quads) (Cont'd)

Region	Manufacturing[a]	Mining	Construction	Total	Total percent
2000					
New England	1.023	0.013	0.095	1.131	2.5
Middle Atlantic	5.100	0.111	0.281	5.492	12.4
East North Central	8.696	0.179	0.314	9.189	20.7
West North Central	1.755	0.248	0.129	2.132	4.8
South Atlantic	4.777	0.201	0.286	5.264	11.8
East South Central	4.046	0.112	0.109	4.267	9.6
West South Central	8.777	1.229	0.172	10.178	22.9
Mountain	1.540	0.557	0.087	2.184	4.9
Pacific	3.919	0.462	0.229	4.610	10.4
U.S. total	39.633	3.112	1.702	44.447	100.0

[a]Includes feedstocks and captive coke.

The Six Major Industries

In Chapter 3 we noted that the six industries listed below [shown with their standard industrial classification (SIC) numbers] account for 85 percent of the total industrial energy demand:

Chemicals and allied products	(28)
Primary metals	(33)
Petroleum and coal products	(29)
Stone, clay, and glass	(32)
Paper and allied products	(26)
Food and kindred products	(20)

In the following section, we discuss the energy estimates for each of the six major manufacturing industries. Our primary source for technological information was *Energy Consumption in Manufacturing* (11). This is followed by a description of the estimates for motor vehicles and other manufacturing, and of mining. In the next section we discuss the adjustment for captive coke and feedstocks. The conversion from net to gross energy is described separately.

The results obtained in this study differ considerably from those of the Federal Power Commission (FPC) (12). Although both the time period covered and their projections of future growth were somewhat different from ours, the chief source of the large disparity was due to their use of fuel price elasticities to determine the reduction in the energy/output ratios in contrast to the IEA reliance on detailed examination of regional trends in each industry. The regional distribution of manufacturing industries that they obtained, assuming an increase in the price of fuels, is shown together with our estimates in Table 5-7. Whereas our estimates show a slight shift from the northern to the southern and from the eastern to the western regions, theirs show a larger movement, generally in the opposite direction.

Chemicals (SIC 28)

To estimate regional energy demand for the chemicals industry, we separated it into three parts. Industrial organic chemicals (SIC 286), which accounted for almost 40 percent of the energy used, are largely derivatives of petroleum and natural gas (13). More than two-thirds of the energy demand for this sector was concentrated in Texas and Louisiana. These two states are expected to remain dominant, although California is likely to increase its share. The energy demand for industrial inorganic chemicals (SIC 281), which accounted

Table 5-7.

Energy in manufacturing, by regional percentage
(Units are Btu's converted to percentages)

Region	IEA		FPC	
	1974	2000	1974	2000
New England	2.4	2.3	2.8	3.6
Middle Atlantic	14.7	12.3	17.2	16.9
East North Central	25.1	21.0	25.1	26.9
West North Central	4.7	4.4	4.3	5.3
South Atlantic	9.9	11.7	11.5	13.5
East South Central	7.1	8.8	7.9	8.6
West South Central	25.9	26.2	20.4	13.1
Mountain	2.6	4.3	2.3	2.5
Pacific	7.6	9.0	8.5	9.6
U.S. total	100.0	100.0	100.0	100.0

for 24 percent of the energy used, is not as geographically clustered. The
major geographic shift we foresee would be linked with the rapid expansion
of uranium diffusion plants. There are no present plans to build new plants;
we therefore assumed that the expansion would occur in the three states that
currently have plants, Ohio, Kentucky, and Tennessee. The third sector is the
rest of the chemicals industry (see below).

The growth rate of the organic chemicals industry increased very rapidly between 1954 and 1967, at about 10 percent per year. Since then, the growth
rate has slowed considerably, to about 5 percent a year. The rapid growth
occurred during a period when prices in this industry declined. With the future expectation of increasing petroleum prices and limited supply, we estimate a 5 percent annual growth to 1985 and a 6 percent annual growth
thereafter. There was a steady small decline in energy used per dollar of ship-
ment (in constant dollars) of 0.7 percent per year between 1954 and 1967.
This value is expected to increase considerably in response to rising energy
prices. New processes, similar to those currently used in other industrialized
countries, and better use of process heat are expected to reduce energy needs

per unit of output by 2.0 percent per year to 1985 and 2.5 percent per year to 2000. Our estimate of the energy requirements for industrial organic chemicals is 1.6 quads in 1985 and 2.6 quads in 2000.

In distributing this demand among regions, we took account of the projected shortage of natural gas, which is expected to shift some of the feedstock requirements from gas products to coal derivatives, and of the importation of methanol. We further assumed that growth in California will increase, because this state is expected to be an energy conserver.

The growth rate of the industrial inorganic chemicals industry increased at an average annual rate of 6.1 percent between 1954 and 1967. Between 1967 and 1974 the value of shipments (measured in constant dollars) actually declined. The most important segment of this industry in terms of energy requirements are uranium diffusion plants. In 1958, these plants accounted for 22 percent of the value of shipments and for 53 percent of the gross energy consumed by the industry. Technological improvements are expected to reduce electricity requirements per unit of output in uranium diffusion plants considerably (in 1967, the gross energy was three times the net energy), but these savings will be more than offset by the rapid increase in production. We have estimated an annual increase in uranium diffusion production of 8 percent per year to 1985, based on Energy Research and Development Administration projections of enrichment capacity at the three existing gaseous diffusion plants. After 1985, enrichment technology is likely to be altered radically and energy requirements could drop dramatically—to perhaps one-tenth of the present requirements per unit. Just which technology will be adopted cannot be predicted. For the purposes of this study, we have assumed that energy requirements for uranium enrichment in 2000 will be the same as in 1985, admittedly a very arbitrary guess. Energy efficiency to 1985 is assumed to increase 3.0 percent per year.

For the rest of the chemicals industry we assumed a growth rate, based largely on our gross national product (GNP) projections, of 4.0 percent to 1985 and 3.0 percent to 2000. Energy per unit of output is expected to decrease as a result of both technological changes and a shift to less energy-intensive chemicals as these become economically less attractive. The reduction in energy per dollar of output is estimated at 2 percent per year to 2000.

The chemicals other than inorganic (SIC 281) and organic (SIC 286) include a large composite of chemical products and are most important in terms of energy use, being plastics materials, synthetics, and fertilizers. These industries have grown rapidly in the past 25 years, although the rate of growth has

slowed down in the past 5 years. The energy/output ratio declined rapidly in
the manufacture of plastics (3.5 percent per year between 1947 and 1967)
and man-made fibers (3.0 percent for cellulose and 2.4 percent for noncellu-
lose between 1958 and 1967), but for synthetic rubber there was an increase
in the energy/output ratio resulting from a change in the product mix. We
have assumed a 6.0 percent annual increase in production to 1985 and a 5.0
percent increase to 2000, accompanied by a 3.0 percent annual decline in the
energy/output ratio to 1985 and 2.5 percent to 2000.

Our projections for organic chemicals, inorganic chemicals, and other chemicals are shown separately in Table 5-8 and are summarized in Table 5-9.

Primary Metals (SIC 33)

The primary metals industries (SIC 33) used 2.64 quads of energy in 1974, or
20 percent of all manufacturing industries. They were the largest users of coal
and electricity, accounting for almost half of the total coal used by industry
and more than one-fourth of the electricity. They also were responsible for
18 percent of the oil and 16 percent of the natural gas used by industry.

The industry is primarily concentrated in four states, Pennsylvania, Ohio,
Indiana, and Illinois. These four states consumed 70 percent of the total energy used for blast furnaces and steel mills, more than half of the energy for
all primary metals industries, and 68 percent of the fuel oil.

The primary metals industries are a composite of a number of major industries whose projected growth, fuel composition, and geographic locations
differ. Table 5-10 divides this industry into its five major components.

The largest industry among the primary metals is blast furnaces and basic
steel products. The average annual growth rate for this industry since 1947
has been 2.0 percent. The Bureau of Economic Analysis projection to 1990
assumes that this growth rate will continue (14). Although IEA projections
anticipate that demand from construction, automobiles, and highways will
grow more slowly in the next two decades than in the past, demand from
mass transit and pipelines is expected to increase. Our estimates assume a 2.0
percent annual increase to 1985 and a 1.8 percent increase from 1985 to
2000. The amount of gross energy required to produce a ton of raw steel declined at an average annual rate of 1.2 percent between 1947 and 1972. With
increasing fuel prices and other incentives for fuel economy, we estimate a
1.5 percent decline to 2000. This would bring the fuel utilization efficiency
in this industry to the current level in West Germany. Our estimate of purchased energy for blast furnaces and basic steel products is 1.7 quads in 1985
and 1.8 quads in 2000.

Table 5-8.
Primary chemical products (SIC 281, 286, and other),
regional distribution of energy requirements, 1974-2000 (x 10^{12} Btu)

Region	Industrial organic	Industrial inorganic	Other	Total
1974				
New England	2.0	0.7	27.0	29.7
Middle Atlantic	90.8	63.1	107.8	261.7
East North Central	60.7	136.1	159.7	356.5
West North Central	3.1	23.5	88.7	115.3
South Atlantic	123.9	76.1	233.7	433.7
East South Central	22.2	123.9	205.7	351.8
West South Central	752.0	202.3	297.9	1252.2
Mountain	—	—	23.2	23.2
Pacific	12.6	32.4	42.0	87.0
U.S. total	1067.3	658.1	1185.7	2911.1
1985				
New England	3.1	1.4	33.1	37.6
Middle Atlantic	132.0	78.5	129.0	339.5
East North Central	88.0	238.2	192.1	518.3
West North Central	4.8	33.4	105.1	143.3
South Atlantic	185.6	108.1	299.9	593.6
East South Central	36.2	218.4	261.0	515.6

Table 5-8.
Primary chemical products (SIC 281, 286, and other),
regional distribution of energy requirements, 1974-2000 (x 10^{12} Btu) (Cont'd)

Region	Industrial organic	Industrial inorganic	Other	Total
1985				
West South Central	1061.8	275.7	378.0	1715.5
Mountain	20.5	1.4	37.5	59.4
Pacific	41.0	42.3	64.5	147.8
U.S. total	1573.0	997.4	1500.2	4070.6
2000				
New England	5.1	1.7	43.3	50.1
Middle Atlantic	216.0	91.1	173.7	480.8
East North Central	145.0	261.0	260.7	666.7
West North Central	7.8	38.9	130.3	177.0
South Atlantic	324.1	125.6	434.3	884.0
East South Central	63.1	230.7	378.0	671.8
West South Central	1712.1	320.0	549.3	2581.4
Mountain	52.5	1.7	87.0	141.2
Pacific	108.2	49.1	123.9	281.2
U.S. total	2633.9	1119.8	2180.5	5934.2

Table 5-9.

Chemicals and allied products (SIC 28),
regional distribution of energy requirements, 1974-2000

Region	Energy (x 10^{12} Btu)			Percent distribution		
	1974	1985	2000	1974	1985	2000
New England	29.7	37.6	50.1	1.0	0.9	0.8
Middle Atlantic	261.7	339.5	480.8	9.0	8.3	8.1
East North Central	356.5	518.3	666.7	12.2	12.7	11.2
West North Central	115.3	143.3	177.0	4.0	3.5	3.0
South Atlantic	433.7	593.6	884.0	14.9	14.6	14.9
East South Central	351.8	515.6	617.8	12.1	12.7	11.3
West South Central	1252.2	1715.5	2581.4	43.0	42.1	43.5
Mountain	23.2	59.4	141.2	0.8	1.5	2.4
Pacific	87.0	147.8	281.2	3.0	3.6	4.7
U.S. total[a]	2911.1	4070.6	5880.2	100.0	100.0	100.0

[a]Totals may not add as a result of rounding.

The growth rate of the iron and steel foundries and forgings industry is estimated to increase at a somewhat slower rate in the future than in the past, as a result of the trend toward smaller and lighter automobile engines and the increasing competition of aluminum. The average growth rate in this industry since 1962 was about 5 percent per year. We have estimated a 3.5 percent growth rate to 2000, based largely on our estimate of the growth rate of automobiles, the trend toward lighter automobile engines, and the increased needs for railroad equipment.

There has been almost no improvement in energy efficiency in the iron and steel foundries and forgings industry over the period from 1947 to 1974. Although there have been considerable energy savings in each of the processes, these have been counteracted by the increasing mechanization in the foundry industries and the installation of pollution-control facilities. The unavailability of gas and the consequent switch to electric furnaces will further counteract anticipated energy efficiencies, since electric furnaces require about 40

Table 5-10.

Major primary metals industries (SIC 33) and energy consumption, 1974

Industry	Energy used (x 10^{12} Btu)	Percent of energy expenditure			
		Electricity	Oil	Coal	Gas
Blast furnaces and basic steel products (SIC 331)	1652.1	31.5	21.3	24.8	17.5
Iron and steel foundries and forgings (SIC 332, 3391, 3399)	187.0	45.3	2.8	30.7	15.0
Primary nonferrous metals (SIC 333, 344)	567.8	67.7	5.7	6.5[a]	17.3
Nonferrous rolling and drawing (SIC 335)	174.7	56.7	11.3	3.0[a]	24.4
Nonferrous foundries (SIC 335, 3398)	59.0	51.3	4.1	6.0[a]	34.2
Total primary metals	2640.6	41.2	15.7	20.0	18.2

[a]Estimated.

times as many Btu's to melt a ton of steel as gas furnaces. We have estimated a decline in the energy efficiency of this industry of 0.4 percent to 1985, and then an increase of 0.5 percent to 2000. Our estimate of energy consumption for this industry is 0.3 quad in 1985 and 0.4 quad in 2000.

Aluminum is the largest energy consumer in the primary nonferrous metals industry, accounting for 70 percent of the energy consumption. The growth in the rate of aluminum production has been extremely rapid in the past, at an average annual rate of 8.2 percent from 1947 to 1972. The largest customers for aluminum are the construction industry (22 percent) and transportation (18 percent). Growth in both these industries, as well as in the consumer durables (10 percent), is expected to slow down in the next 25 years. Furthermore, state and county legislation restricting the manufacture of disposable containers (11 percent) will sharply reduce the growth in the demand for aluminum. On the other hand, the emphasis on lighter cars and trains is expected to increase the purchases of aluminum by these industries, despite their slower growth. We have estimated a 5 percent annual growth rate for aluminum to 2000.

The largest opportunity for increased fuel efficiency is a shift from the Hall-Heroult process to the Alcoa process. The latter requires about 5 kilowatt-hours per pound of aluminum, compared to 8 kilowatt-hours per pound for the former. Another important potential for saving energy is increased use of scrap. A pound of a semifabricated aluminum product made from scrap uses only one-fourth the energy of one made from bauxite. Finally, there is the possibility that international firms will shift the production of alumina from the United States to the bauxite-producing countries.

Pressures to increase fuel efficiency in aluminum production as fuel prices increase will be very strong. In 1974, 16 percent of the value of aluminum shipments represented the cost of energy, mostly electricity, which was purchased at half the price paid by other industries. Although the special contract prices for electricity for the aluminum companies will for the most part continue through 1985, new plants, as well as old plant replacements, will be built in a very different economic setting. Electricity in the Pacific area will no longer be so cheap. Furthermore, gas, which provided more than 80 percent of all heat in the fabrication process, is likely to be replaced by other fuels.

We estimate that by 1985 most of the increased energy efficiency in aluminum will result from reducing the wasted heat and increasing the amount of scrap used. These two factors could increase efficiency by about 10 percent. After 1985, technological changes are likely to dominate, with a further energy reduction of perhaps 20 percent.

Our estimate of energy requirements for the nonferrous metals industries are based on a 4.5 percent annual growth rate (the growth rate of copper being less rapid than that of aluminum), with an annual increase in energy efficiency of 1.0 percent to 1985 and 1.2 percent to 2000. Energy consumption for these industries is therefore estimated at 1.2 quads in 1985 and 1.9 quads in 2000. By 2000, the energy requirements for the nonferrous metals are expected to be 46 percent of the total primary metals industry compared to 30 percent in 1974.

Our projections for ferrous and nonferrous metals are shown separately in Table 5-11 and are summarized in Table 5-12.

Table 5-11.

Ferrous and nonferrous metals industries,
regional distribution of energy requirements, 1974-2000
$(x \ 10^6 \ Btu)$

Region	Ferrous metals	Nonferrous metals	Total
1974			
New England	9.2	17.4	26.6
Middle Atlantic	482.8	98.6	581.4
East North Central	901.5	152.5	1054.0
West North Central	8.5	38.2	46.7
South Atlantic	73.0	109.2	182.2
East South Central	152.2	53.9	206.1
West South Central	96.6	169.6	266.2
Mountain	17.7	55.3	73.0
Pacific	50.5	117.4	167.9
U.S. total	1792.0	812.1	2604.1
1985			
New England	11.9	24.6	36.5
Middle Atlantic	523.3	134.4	657.7
East North Central	978.9	210.5	1189.4
West North Central	13.0	60.7	73.7
South Atlantic	91.4	149.8	241.2
East South Central	187.0	79.5	266.5
West South Central	113.6	247.0	360.6
Mountain	44.7	91.4	136.1
Pacific	59.0	172.0	231.0
U.S. total	2022.8	1169.9	3192.7
2000			
New England	13.6	37.5	51.1
Middle Atlantic	573.2	191.1	764.3
East North Central	1030.4	324.1	1354.5

Region	Ferrous metals	Nonferrous metals	Total
2000			
West North Central	17.1	112.6	129.7
South Atlantic	112.6	228.6	341.2
East South Central	225.2	133.1	358.3
West South Central	136.8	419.7	556.5
Mountain	92.1	170.6	262.7
Pacific	68.2	286.6	354.8
U.S. total	2269.2	1903.9	4173.1

Table 5-12.
Primary metals industry (SIC 33),
regional distribution of energy requirements, 1974-2000

Region	Energy (x 10^{12} Btu)			Percent distribution		
	1974	1985	2000	1974	1985	2000
New England	26.6	36.5	51.1	1.0	1.1	1.2
Middle Atlantic	581.4	657.7	764.3	22.3	20.8	18.3
East North Central	1054.0	1189.4	1354.5	40.5	37.1	32.5
West North Central	46.7	73.7	129.7	1.8	2.3	3.1
South Atlantic	182.2	241.2	341.2	7.0	7.5	8.2
East South Central	206.1	266.5	358.3	7.9	8.3	8.6
West South Central	266.2	360.6	556.2	10.2	11.3	13.5
Mountain	73.0	136.1	262.7	2.8	4.3	6.3
Pacific	167.9	231.0	354.8	6.4	7.2	8.5
U.S. total[a]	2604.1	3192.7	4172.8	100.0	100.0	100.0

[a]Totals may not add as a result of rounding.

Petroleum and Coal Products (SIC 29)

The petroleum and coal products industries are concentrated in Texas (46 percent), Louisiana (12 percent), and California (11 percent). We expect Texas and Louisiana to remain dominant, with petroleum from the outer continental shelf and imports from Mexico and the Middle East replacing onshore production. California, which could increase its share, is considered unlikely to do so for environmental reasons.

Growth in this industry is expected to diminish as supplies shrink and prices rise. Between 1954 and 1974, the average annual growth rate was 3.7 percent, with a 2.3 percent annual growth between 1971 and 1974. Between 1974 and 1975 there was no growth; however, there was a 6.8 percent increase in 1976. Our estimate of a 1.0 percent annual growth to 2000 is based on the assumption that gasoline consumption will be sharply curtailed.

The energy requirements of this industry increased more rapidly than production from 1954 to 1962, increased at the same rate from 1963 to 1971, and thereafter rose less. If sales value is used instead of the production index (the quantity produced)—the difference being that the former reflects qualitative differences although the latter does not—the period since 1962 was one of decreasing rather than increasing energy/output ratio (15). Between 1971 and 1974, the energy/production rate declined by 9 percent. We estimate that technological changes, which will greatly reduce waste heat, will result in a further decline in the energy/production ratio of 15 percent by 1985 and an additional 20 percent to 2000.

Our estimates of regional energy demand by the petroleum and coal industries are shown in Table 5-13.

Stone, Clay, and Glass Products (SIC 32).

The stone, clay, and glass products industry is the most energy-intensive industry. Hydraulic cement, which uses 37 percent of the energy of this industry and accounts for only 8 percent of the value of shipments, pays 22 cents out of each dollar of sales for energy. As would be expected, in regions where energy is cheap, such as the West South Central, West North Central, and East South Central, a larger portion of the manufacturing effort is concentrated in the cement industry than in areas of high-energy cost, such as New England. However, the industry is also market-oriented. We anticipate some shift to the West South Central, Mountain, and Pacific regions, in response to both energy price and population pressures.

Growth in this industry has lagged somewhat behind growth in manufactur-

Table 5-13.

Petroleum and coal products (SIC 29),
regional distribution of energy requirements, 1974-2000

Region	Energy (x 10^{12} Btu)			Percent distribution		
	1974	1985	2000	1974	1985	2000
New England	4.1	4.1	3.8	0.3	0.3	0.3
Middle Atlantic	85.0	82.6	79.8	5.7	5.4	5.4
East North Central	134.4	130.3	126.2	9.0	8.6	8.6
West North Central	64.8	62.8	60.7	4.3	4.1	4.1
South Atlantic	18.1	17.4	17.1	1.2	1.1	1.2
East South Central	13.0	12.6	12.3	0.9	0.8	0.8
West South Central	976.5	1008.6	968.7	65.2	66.3	65.8
Mountain	19.4	18.8	18.4	1.3	1.2	1.3
Pacific	182.9	183.6	184.2	12.2	12.1	12.5
U.S. total	1498.2	1520.8	1471.2	100.0	100.0	100.0

ing, 3.6 percent compared to 4.5 percent annually between 1954 and 1974. This lag was due largely to the somewhat slower growth of the construction industry, a shift from exporting to importing cement, and the substitution of other materials for glass and bricks. We anticipate a 2.5 percent growth rate in this industry to 2000, based mainly on our projections of a 1.5 percent annual increase in the number of new housing units (houses, apartments, and trailers) built.

The energy/output ratio declined at an average annual rate of 1.6 percent between 1954 and 1975. However, between 1971 and 1974, it declined at an average annual rate of 3.9 percent. This rapid reduction was due largely to technological improvements, increasing plant and furnace size, and better use of waste heat. These factors, and a shift from the wet process to the less energy-intensive dry process in cement manufacturing, are likely to result in further reductions in the energy/output ratio. We estimate the average annual decline to 1985 as 3.5 percent and to 2000 as 3.2 percent, resulting in a reduction of energy per unit of output of more than half by 2000.

Our estimates of the energy requirements for this industry in 1985 and 2000 by region are shown in Table 5-14.

Table 5-14.

Stone, clay, and glass products (SIC 32),
regional distribution of energy requirements, 1974-2000

Region	Energy (x 10^{12} Btu)			Percent distribution		
	1974	1985	2000	1974	1985	2000
New England	21.6	16.0	10.9	1.6	1.3	1.0
Middle Atlantic	226.6	196.9	153.9	17.1	16.0	14.0
East North Central	306.4	270.6	220.1	23.1	22.0	20.0
West North Central	138.2	115.7	88.0	10.4	9.4	8.0
South Atlantic	188.0	182.2	176.1	14.2	14.8	16.0
East South Central	106.5	101.0	99.0	8.0	8.2	9.0
West South Central	161.0	153.9	143.0	12.1	12.5	13.0
Mountain	56.3	71.3	88.0	4.2	5.8	8.0
Pacific	121.8	123.2	121.1	9.2	10.0	11.0
U.S. total	1326.4	1230.8	1100.1	100.0	100.0	100.0

Paper and Allied Products (SIC 26)

The paper and allied products industry is partially market-located (non-integrated paper and board mills) and partially resource-located (pulp and integrated mills). In 1974, the South Atlantic region was the largest energy consumer (22.8 percent), and the next largest was the East North Central region (19.4 percent). The value of shipments of the East North Central region was about 50 percent greater than that of the South Atlantic region. The discrepancy is due to the preponderance of paperboard mills in the South Atlantic region and of paper mills in the East North Central. Paperboard mills are particularly energy-intensive, the cost of energy being 12.4 percent of the value of shipments in 1974. For paper mills (excluding building paper) the cost of energy was 8.5 percent.

The largest regional shift in this industry is likely to be a growth in the Pacific region. This region accounts for 34 percent of the lumber industry but only slightly more than 10 percent of the paper industry. We also expect some shift toward the Mountain area. The New England, Middle Atlantic, and East North Central regions are expected to decline in importance. These projected shifts assume that large integrated mills, which tend to locate in the timber regions, will supplant the nonintegrated mills. This shift seems particularly likely in view of increasing energy prices. The integrated mills use waste products for logging and pulping, which the nonintegrated mills do not have.

Production in paper has generally grown slightly faster than the GNP. We assume that this trend will continue, with a somewhat slower growth for paper and paperboard mills (where higher prices will inhibit growth somewhat) and a faster growth in the other sectors (due largely to increased demand for insulation materials). We have assumed a 4.5 percent growth to 1985 and a 3.5 percent growth to 2000.

The energy/output ratio for the paper industry declined by 4 percent each year between 1971 and 1974, considerably more than the average annual decline of 1.3 percent for the period 1954 to 1974. In recent years, the industry has invested heavily in modernizing its plants to meet environmental standards. Further investment to increase energy efficiency is likely because energy costs are such a high portion of the total costs. Furthermore, since oil has been the primary fuel of this industry (52 percent of the energy costs was for oil in paperboard mills and 38 percent in paper mills), such investment seems inevitable. We estimate that energy consumption per unit of output in 1985 will be 80 percent of that in 1974 and will be 60 percent in 2000. Our estimates of purchased energy requirements for the paper industry, by region, are shown in Table 5-15.

Food and Kindred Products (SIC 20)

The food and kindred products industry is concentrated in the agricultural regions. Illinois and California accounted for one-fifth of the total energy used by this industry in 1974. Wisconsin, Minnesota, and Iowa used one-sixth. We anticipate only a minor regional shift based on population movement toward the South, and on a continuation of the trend toward more meat consumption.

Production in the food industry grew at an average annual rate of 3.6 percent between 1954 and 1974. Because of the slower population growth projected for the next decades, we have estimated an annual growth rate of 3.2 percent to 2000.

The annual energy/output ratio dropped rapidly between 1954 and 1958 (4 percent) and in 1971 through 1974 (5 percent), although it declined only slightly (1 percent) in the intervening years. In the meat-packing plants, energy savings were realized largely as a result of the trend toward heavier

Table 5-15.

Paper and allied products (SIC 26),
regional distribution of energy requirements, 1974-2000

Region	Energy (x 10^{12} Btu)			Percent distribution		
	1974	1985	2000	1974	1985	2000
New England	117.4	177.1	190.0	9.0	8.2	7.0
Middle Atlantic	148.8	224.5	244.3	11.5	10.4	9.0
East North Central	252.1	399.5	461.3	19.4	18.5	17.0
West North Central	43.3	75.7	108.5	3.3	3.5	4.0
South Atlantic	295.8	466.4	542.5	22.8	21.6	20.0
East South Central	143.3	216.0	244.3	11.0	10.0	9.0
West South Central	150.5	237.5	271.3	11.6	11.0	10.0
Mountain	1.0	47.4	108.5	0.1	2.0	4.0
Pacific	145.0	323.8	542.5	11.2	15.0	20.0
U.S. total[a]	1297.2	2167.9	2713.2	100.0	100.0	100.0

[a]Because of rounding, totals may not add.

animals. In fluid milk production, the shift from bottles to plastic-lined cartons and to large containers resulted in considerable energy savings. For frozen fruits and vegetables, the energy/output ratio increased, largely as a result of a shift from slow to quick freezing. In the bread and cake industry, there was some energy saving resulting from the use of larger, more efficient ovens.

Although the food industry is a large user of energy, energy costs are a small portion of the total costs. Price increases of energy thus will not provide a strong impetus for energy savings, although they will have some impact. The trend toward increased meat consumption by comparison with other foods is likely to lead to some reduction in the energy/output ratio, since meat-packing uses about half the energy per dollar of value of the food industry as a whole. We have assumed that the energy/output ratio will decline at an average annual rate of 2.5 percent to 1985 and 2.0 percent to 2000. Our estimates of energy consumption in the food industry are shown in Table 5-16.

Table 5-16.

Food and kindred products (SIC 20),
regional distribution of energy requirements, 1974-2000

Region	Energy (x 10^{12} Btu)			Percent distribution		
	1974	1985	2000	1974	1985	2000
New England	23.9	24.6	28.0	2.5	2.4	2.3
Middle Atlantic	112.9	120.8	139.6	12.0	11.8	11.5
East North Central	233.7	256.9	307.8	24.9	25.1	25.4
West North Central	164.8	177.1	206.1	17.5	17.3	17.0
South Atlantic	94.2	103.4	123.5	10.0	10.1	10.2
East South Central	46.4	49.1	54.6	4.9	4.8	4.5
West South Central	81.2	88.0	105.4	8.6	8.6	8.7
Mountain	58.0	67.6	85.0	6.2	6.6	7.0
Pacific	124.5	136.1	162.4	13.3	13.3	13.4
U.S. total	939.6	1023.5	1212.4	100.0	100.0	100.0

Projections of Other Industries

As noted in Chapter 3, the six major industries that we have just considered
account for 80 percent of the energy consumed in manufacturing. Since other
industries account for relatively little of the energy use, they are projected
here as a group, except for the auto industry. Since our 101-quad scenario an-
ticipates a secular slowdown in auto production, the energy consumption
forecast of that industry is calculated separately (see Chapter 1). The national
control totals are distributed regionally as they were in 1974 because of the
heterogeneous nature of this industry group.

The percentage of energy used in 1974 and projected for 2000 by industry,
which is accounted for by the six major energy-consuming industries plus the
auto industry, is shown in Table 5-17. Table 5-18 projects regional energy use
by the six major energy consumers and the auto and other manufacturing industries for 1985 and 2000.

Table 5-17.

Percent of energy used by the six
largest industrial consumers plus the
auto industry, 1974-2000, in each
geographic region

Region	1974	2000
New England	56.2	59.6
Middle Atlantic	77.5	77.5
East North Central	79.7	80.0
West North Central	79.6	79.9
South Atlantic	80.6	85.0
East South Central	80.8	83.9
West South Central	94.8	95.7
Mountain	64.1	79.6
Pacific	79.1	85.0
U.S.	80.7	84.4

Gross Energy Demand

The projections of energy requirements for manufacturing given in Tables
5-17 and 5-18 are net requirements, that is, energy delivered to the industry.
Gross requirements include the energy lost in electrical generation and transmission. In Chapter 3 the gross demand for manufacturing was projected on
the basis of the general U.S. trend toward electrification, modified to take
account of special needs in each of the six major energy-using industries. Regional industrial electrification is calculated by projecting the overall trends
to each region. These overall projections were modified for two regions where
the choice between the direct burning of coal versus the use of electricity
probably will diverge from the national experience. The regions were New
England and Mountain, where coal is, respectively, most expensive and least
expensive. We made 5 percent changes in each of the two extreme regions to
account for this possibility (see Table 5-19).

Table 5-18.
Regional distribution of energy consumption[a] by industry groups, 1974-2000 (x 10^{12} Btu)

Region	Six major industries	Auto	Other	Total manufacturing
1974				
New England	222.1	1.4	173.0	396.5
Middle Atlantic	1,416.3	18.1	416.3	1,850.7
East North Central	2,340.3	184.2	644.9	3,169.4
West North Central	573.2	11.6	149.4	734.2
South Atlantic	1,201.0	0.7	276.7	1,478.4
East South Central	869.7	6.5	210.5	1,086.7
West South Central	2,889.3	5.8	152.5	3,047.6
Mountain	231.0	0.3	128.6	359.9
Pacific	831.2	6.1	223.5	1,060.8
U.S. total	10,574.1	234.7	2375.4	13,184.2
1985				
New England	296.2	1.4	211.9	509.5
Middle Atlantic	1,630.9	22.2	521.7	2,174.8
East North Central	2,765.1	226.5	856.1	3,847.7
West North Central	648.3	14.3	189.0	851.6
South Atlantic	1,604.3	0.7	343.2	1,948.2
East South Central	1,160.8	8.2	263.1	1,432.1

Table 5-18.
Regional distribution of energy consumption[a] by industry groups, 1974-2000 (x 10^{12} Btu) (Cont'd)

Region	Six major industries	Auto	Other	Total manufacturing
1985				
West South Central	3,564.2	7.2	190.7	3,762.1
Mountain	400.6	0.3	159.3	560.2
Pacific	1,145.4	7.5	276.0	1,428.9
U.S. total	13,215.8	288.3	3011.0	16,515.1
2000				
New England	334.0	1.4	227.9	563.3
Middle Atlantic	1,862.6	16.0	551.0	2,429.6
East North Central	3,136.7	165.5	875.2	4,177.4
West North Central	770.1	10.2	201.0	981.3
South Atlantic	2,084.4	0.7	364.7	2,449.8
East South Central	1,440.2	5.5	278.1	1,723.8
West South Central	4,626.3	5.1	202.0	4,833.4
Mountain	703.9	0.3	169.6	873.8
Pacific	1,646.3	5.5	293.1	1,944.9
U.S. total	16,604.5	210.2	3162.6	19,977.3

[a]Excludes captive coke and feedstocks.

Table 5-19.

Net and gross energy requirements, by region,
for total manufacturing, 1974-2000
$(x\ 10^{12}\ Btu)$

Region	Net energy	Gross electricity	Gross energy
1974			
New England	396.5	213.3	538.7
Middle Atlantic	1,850.7	922.3	2,465.6
East North Central	3,169.4	1575.0	4,219.4
West North Central	734.2	293.1	929.6
South Atlantic	1,478.4	838.0	2,037.1
East South Central	1,086.7	897.7	1,685.2
West South Central	3,047.6	762.2	3,555.7
Mountain	359.9	192.1	488.0
Pacific	1,060.8	807.0	1,598.8
U.S. total	13,184.2	6500.7	17,518.1
1985			
New England	509.5	356.2	747.0
Middle Atlantic	2,174.8	1,483.9	3,164.0
East North Central	3,847.7	2,711.5	5,655.4
West North Central	851.6	474.3	1,167.8
South Atlantic	1,948.2	1,344.7	2,844.7
East South Central	1,432.1	1,455.2	2,402.2
West South Central	3,762.1	1,224.6	4,578.5
Mountain	560.2	294.7	756.7
Pacific	1,428.9	1,299.6	2,295.3
U.S. total	16,515.1	10,644.7	23,611.6
2000			
New England	563.3	571.5	944.3
Middle Atlantic	2,429.6	2,381.9	4,017.5

Table 5-19.

Net and gross energy requirements, by region,
for total manufacturing, 1974-2000
$(x \ 10^{12} \ Btu)$ (Cont'd)

Region	Net energy	Gross electricity	Gross energy
2000			
East North Central	4,177.4	4,076.7	6,895.2
West North Central	981.3	759.9	1,487.9
South Atlantic	2,449.8	2,154.7	3,886.2
East South Central	1,723.8	2,331.4	3,278.1
West South Central	4,833.4	1,963.6	6,142.5
Mountain	873.8	478.7	1,192.9
Pacific	1,944.9	2,045.8	3,308.8
U.S. total	19,977.3	16,764.2	31,153.4

Feedstocks and Captive Coke

Captive coke is of energy importance in the primary metals industry, and
feedstocks are a factor in the petroleum and chemical industries. Growth in
these consumption areas are forecast at the same rate as that projected for the
industries that use fuels for their energy content. The regional dynamics of
these industries are quite different. The petroleum industry is expected to
grow only slowly; this means that the bulk of the feedstock growth will shift
from petroleum to chemicals; by 2000, only 16 percent of feedstocks are estimated to be captive to petroleum output (see Table 5-20).

Mining

The oil- and gas-extraction industry is the single largest consumer of energy in
mining, accounting for two-thirds of the total gross energy used in 1972 (1.53
quads). This explains why half the energy consumed is in the West South Central region. The Mountain and Pacific regions, which ranked second and third
in energy consumption in mining, were heavy producers of petroleum. The
next largest energy consumer at 0.37 quad was nonmetallic mining, then
metallic mining at 0.25 quad, and finally coal mining at 0.15 quad.

The most important development anticipated in the mining industry between 1975 and 2000 is a large increase in coal production and a small decrease in oil and gas production. This development implies regional shifts in
the energy required for the mining sector.

The production of crude oil is expected to decline slightly from the 1975
level of 8.4 billion barrels a day to 8.3 billion barrels a day in 2000. The West
South Central region, which now supplies two-thirds of this oil, is expected to
supply only one-half by 2000, with Alaska supplying one-fourth. Likewise,
we anticipate a small decrease in natural gas production, from 19.9 to 18.4
quads. The West South Central region, which accounted for 81 percent of
this production in 1975, is expected to produce 75 percent in 2000. Alaska,
which in 1975 produced less than 1 percent of the gas, is expected to produce
8 percent in 2000. The West South Central region is therefore expected to
decrease its share of energy used in mining as a result of both the shift from
gas and oil to coal and the future importance of Alaska as a source of gas and
oil. The projections for oil and natural gas are shown in Tables 5-21 and 5-22.

Metal mining required 253.2×10^{12} Btu in 1972 and nonmetallic mining
(not including fuels) 365.1×10^{12} Btu. The Mountain region produced most
of the copper, whereas most of the iron ore was mined in Minnesota, Michigan, California, and Wyoming. For nonmetallic minerals, the leading states
were California, Texas, and Michigan for portland cement, sand, and gravel;
Pennsylvania for cement and stone; and Illinois for sand, gravel, and stone.

Our estimates of the regional distribution for nonfuel mining was based on
the value added by each region for metal and for nonmetal mining in 1972.
For metal mining our estimate of production growth is 1.4 percent a year,
based on the growth rate of the past two decades. For nonmetal mining, production is estimated to increase by 2.5 percent per year.

Table 5-23 shows the regional distribution of gross energy for mining in
1974, plus our projections for 2000. Table 5-24 shows the estimates for energy requirements for each major group of materials mined by region in 2000.

Construction

Growth in the construction industry is projected to correlate with the estimated growth in the numbers of housing units. Our projection of housing
units for 2000 is 102×10^6, up from 72×10^6 in 1975, an increase of 42 percent. We do not project any notable increases in the use of energy by the con-

Table 5-20.

Feedstocks and captive coke, fuels, and total gross energy for manufacturing, by geographic region, 1974–2000 (x 10^{12} Btu)

Region	Feedstocks	Captive coke	Fuels	Total gross energy
1974				
New England	0.0	30.0	538.7	568.7
Middle Atlantic	160.0	610.0	2,465.6	3,235.6
East North Central	180.2	1121.9	4,219.4	5,521.5
West North Central	69.9	39.9	929.6	1,039.4
South Atlantic	109.9	170.3	2,037.1	2,317.3
East South Central	30.0	150.1	1,685.2	1,865.3
West South Central	1576.3	269.9	3,555.7	5,401.9
Mountain	30.0	79.8	488.0	597.8
Pacific	200.0	90.1	1,598.8	1,888.9
U.S. total	2356.3	2562.0	17,518.1	22,436.4
1985				
New England	19.8	34.1	747.0	800.9
Middle Atlantic	224.5	644.5	3,164.1	4,033.1
East North Central	346.7	1150.2	5,655.4	7,152.3

West North Central	109.2	71.3	1,167.8	1,348.3
South Atlantic	329.6	232.4	2,844.7	3,406.7
East South Central	285.6	257.3	2,402.2	2,945.1
West South Central	1449.1	350.1	4,578.5	6,377.7
Mountain	720.0	137.2	756.7	1,613.9
Pacific	173.3	223.1	2,295.3	2,691.7
U.S. total	3657.8	3100.2	23,611.7	30,369.7

2000

New England	29.3	48.5	944.3	1,022.1
Middle Atlantic	337.8	744.5	4,017.5	5,099.8
East North Central	478.0	1322.8	6,895.2	8,696.0
West North Central	140.6	126.2	1,487.9	1,754.7
South Atlantic	556.5	333.7	3,886.2	4,776.4
East South Central	421.7	346.0	3,278.1	4,045.8
West South Central	2093.6	541.1	6,142.5	8,777.2
Mountain	88.0	259.0	1,192.9	1,539.9
Pacific	264.4	346.0	3,308.8	3,919.2
U.S. total	4409.9	4067.8	31,153.4	39,631.1

Table 5-21.

Crude oil production by regions, 1975 and 2000

Region	1975 (quads)	2000 (quads)
New England		
Middle Atlantic	0.02	0.01
East North Central	0.42	0.20
West North Central	0.56	0.25
South Atlantic	0.30	0.50
East South Central	0.44	0.30
West South Central	13.35	10.00
Mountain	2.25	2.00
Pacific (excluding Alaska)	2.01	1.50
Alaska	0.53	4.70
U.S. total	19.88	19.46

struction industry or regional shifts. These assumptions generate a national projection of 1.7 quads in 2000 with the regional breakdown shown in Table 5-25.

Industrial Summary

The summary and analysis of industrial demand for 1974 and 2000 were given in Table 5-6.

Electric Power

The national electric power share of total energy in 2000, as estimated by IEA (see Chapter 4), is about 46 percent. The latest regional breakdown from the BOM (16) shows four regions whose ratio of electric power consumption to total energy use is now about equal to the national average. Five regions, accounting for about 50 percent of the total energy consumption, showed markedly different ratios of electricity to total energy. The percentages for these regions were adjusted for their current percentage deviation from the national average and projected to 2000 (see Table 5-26). These percentages, applied to the regional total energy estimates derived as explained above, then

Table 5-22.

Natural gas production by regions, 1975 and 2000

Region	1975		2000	
	Quads	Percent	Quads	Percent
New England	—	—	—	—
Middle Atlantic	0.09	0.5	0.08	0.4
East North Central	0.19	1.0	0.18	1.0
West North Central	0.85	4.3	0.74	4.0
South Atlantic	0.20	1.0	0.17	0.9
East South Central	0.18	0.9	0.15	0.8
West South Central	16.13	81.0	13.76	74.8
Mountain	1.78	8.9	1.55	8.4
Pacific (excluding Alaska)	0.32	1.6	0.27	1.5
Alaska	0.16	0.8	1.50	8.2
U.S. total	19.90	100.0	18.40	100.0

gave a gross electrical energy figure for each region.

From Federal Power Commission data (17), a regional breakdown of 1975 electric power generation in Btu's by source was obtained—coal, petroleum, natural gas, hydro power, nuclear power, and other. To the 1975 data were added growth estimates based on the planned state-by-state utility expansions identified by the BOM in 1976 (18). Recaps are given in Tables 5-27 and 5-28. These plant capacities were then added to current totals to give the electricity generation by type of fuel for 2000. In several regions the additions to electric power capacity, although scheduled for completion by 1985, were enough or were greater than needed, on the basis of IEA estimates, to meet the estimated electric power requirement for 2000; in other regions they were not. Therefore, modifications were made to bring projected supply into line with estimated demand.

Regional worksheets were prepared giving plant-by-plant data from the BOM publications (18). In writing planned regional expansions to the total calculated electric power required in 2000, the following rules governed:

1. The plants projected for the 1990s were deleted, since their construction was least certain.

Table 5-23.

Energy use in mining, by region, 1974 and 2000
(x 10^{12} Btu)

Region	Gross energy	
	1974	2000
New England	4.6	12.9
Middle Atlantic	68.1	110.9
East North Central	118.8	180.0
West North Central	161.3	248.5
South Atlantic	148.6	200.8
East South Central	85.8	111.9
West South Central	1156.3	1229.4
Mountain	326.0	556.7
Pacific	279.1	461.7
U.S. total	2348.6	3112.8

Source: 1972 regional distribution of purchased fuels and electricity plus energy produced and consumed, Census of Minerals Industries, "Fuels and Electric Energy Consumed," Publication MIC 72(1)-5, Bureau of the Census, Washington, D.C., November 1975, Table V. The 1974 projections were made by IEA.

2. If there was significant electric power generation fired by oil or natural gas planned in a region where it would probably be replaced by coal or nuclear power, because of federal restrictions on new construction of these facilities, oil and natural gas capacity additions were cut back.

3. In increasing planned capacity to meet IEA estimated regional needs where the BOM projections appeared low, the general rule was to add on the same fuel mix in proportion to what was already planned for generating expansion in the region. This meant that expansions were largely coal or nuclear-powered additions to plans.

The object of this exercise was to estimate for 2000 a percentage breakdown of generating capacity by fuel type, by the use of which gross electric power generation in 2000 could be allocated by individual fuel types. No assumptions with respect to operating efficiency were made, since the goal was

Table 5-24.

Estimated gross energy in mining, 2000
(x 10^{12} Btu)

Region	Coal		Oil and gas		Metal		Nonmetal	
	Btu	Percent	Btu	Percent	Btu	Percent	Btu	Percent
New England	—	—	—	—	4.2	1.0	8.7	1.5
Middle Atlantic	19.9	6.6	3.6	0.2	21.0	5.0	66.3	11.4
East North Central	32.6	10.8	18.1	1.0	31.2	7.4	97.1	16.7
West North Central	49.5	16.4	47.0	2.6	110.7	26.3	41.3	7.1
South Atlantic	36.5	12.1	30.7	1.7	3.4	0.8	130.2	22.4
East South Central	44.4	14.7	20.0	1.1	6.3	1.5	41.3	7.1
West South Central	0.3	0.1	1149.1	63.6	8.4	2.0	71.5	12.3
Mountain	117.8	39.0	169.8	9.4	215.5	51.2	53.4	9.2
Pacific	1.4	0.5	368.6	20.4	20.2	4.8	71.5	12.3
U.S. total[a]	302.4	100.0	1806.9	100.0	420.9	100.0	581.3	100.0

[a]Because of rounding, totals may not add.

Table 5-25.

Regional construction energy demand, 1974 and 2000
$(x\ 10^{12}\ Btu)$

Region	1974	2000
New England	67.9	95.2
Middle Atlantic	217.3	280.8
East North Central	229.3	314.2
West North Central	95.5	128.6
South Atlantic	182.5	285.9
East South Central	73.0	108.8
West South Central	112.9	172.3
Mountain	50.5	87.0
Pacific	163.4	228.9
U.S. total	1192.3	1701.7

not to derive a "capacity-required" estimate, which would have provided a 20 to 25 percent reserve. The results are shown in Table 5-29. On the basis of these calculations, coal is expected to account for about the same share of power generation in 2000 as in 1975, whereas the relative contribution of oil and gas will be cut by one-half and two-thirds, respectively, and the nuclear power contribution will be roughly triple the present value.

Recapitulation of Estimates

The estimated increase in energy demand between 1975 and 2000 for the nation as a whole is given in Table 5-30. It can be seen that, although all consuming sectors are expected to grow, the largest increase is anticipated to be in industry. Moreover, the contribution of electricity to the total gross energy is expected to increase to about 46 percent. Regional breakdowns are given in Table 5-31.

Table 5-26.

Electricity share of total regional energy consumption, 2000

Electricity/total energy ratio	Percent
National average	46.6
New England, Middle Atlantic, East North Central,	
and West North Central regions	46.6
South Atlantic region	56.6
East South Central region	58.6
West South Central region	40.6
Mountain region	52.6
Pacific region	55.6

Table 5-27.

Electric power expansion projects in the eastern states as of June 1976 (megawatts)

Region	Total	Coal	Oil/gas	Hydro power	Nuclear power	Geothermal energy	Unknown or other
New England							
Total	13,747	—	2,347	830	10,380	—	190
Maine	2,580	—	600	830	1,150	—	—
New Hampshire	2,300	—	—	—	2,300	—	—
Vermont	369	—	329	—	—	—	40
Massachusetts	5,034	—	1,404	—	3,480	—	150
Rhode Island	2,314	—	14	—	2,300	—	—
Connecticut	1,150	—	—	—	1,150	—	—
Middle Atlantic							
Total	53,412	6,487	4,617	7,800	34,366	—	142
New York	27,701	3,250	2,050	4,000	18,369	—	32
Pennsylvania	11,541	3,237	816	1,500	5,988	—	—
New Jersey	14,170	—	1,751	2,300	10,009	—	110

South Atlantic							
Total	75,856	14,647	11,616	5,097	43,846	—	650
Delaware	2,340	800	—	—	1,540	—	—
Maryland	17,653	1,480	4,567	1,000	9,956	—	650
West Virginia	3,552	2,552	—	1,000	—	—	—
Virginia	6,478	—	—	1,150	5,328	—	—
North Carolina	12,795	2,160	14	—	10,621	—	—
South Carolina	8,086	560	—	480	7,046	—	—
Georgia	10,781	5,324	850	1,467	3,140	—	—
Florida	14,171	1,771	6,185	—	6,215	—	—
East South Central							
Total	44,030	13,059	207	2,061	28,703	—	—
Alabama	16,306	5,700	—	270	10,336	—	—
Mississippi	3,900	1,400	—	1,730	2,500	—	—
Tennessee	17,597	—	—	61	15,867	—	—
Kentucky	6,227	5,959	207	—	—	—	—

Table 5-27.
Electric power expansion projects in the eastern states as of June 1976
(megawatts) (Cont'd)

Region	Total	Coal	Oil/gas	Hydro power	Nuclear power	Geothermal energy	Unknown or other
East North Central							
Total	53,899	21,751	4,703	30	27,415	—	—
Ohio	11,213	4,105	—	—	7,108	—	—
Indiana	11,655	8,750	—	—	2,905	—	—
Illinois	15,208	4,145	2,500	—	8,563	—	—
Michigan	10,246	2,594	1,763	—	5,889	—	—
Wisconsin	5,577	2,157	440	30	2,950	—	—
East of Mississippi							
Totals	240,944	55,944	23,490	15,818	144,710	—	982
Percent	100.0	23.2	9.7	6.6	60.0	—	0.4

Table 5-28.

Electric power expansion projects in the western states as of May 1976 (megawatts)

Region	Total	Coal	Oil/gas	Hydro power	Nuclear power	Geothermal energy	Unknown or other
West North Central							
Total	30,093.5	19,942	3,155.5	4,700	2,250	—	46
Minnesota	4,308	4,304	4	—	—	—	—
Iowa	4,406	2,581	725	—	1,100	—	—
Missouri	7,100	2,655	1,895	2,550	—	—	—
Kansas	5,936	4,275	465	—	1,150	—	46
Nebraska	4,217	2,067	—	2,150	—	—	—
South Dakota	1,306.5	1,240	66.5	—	—	—	—
North Dakota	2,820	2,820	—	—	—	—	—
West South Central							
Total	47,276	30,769	4,046	—	11,641	—	820
Oklahoma	6,891	4,600	1,341	—	950	—	—
Arkansas	6,243	5,293	—	—	950	—	—
Louisiana	6,455	3,180	230	—	3,045	—	—
Texas	27,687	17,696	2,475	—	6,696	—	820
Mountain							
Total	36,990	30,575	431	2,174	3,810	—	—

Table 5-28.
Electric power expansion projects in the western states as of May 1976 (megawatts) (Cont'd)

Region	Total	Coal	Oil/gas	Hydro power	Nuclear power	Geothermal energy	Unknown or other
Montana	4,397	2,980	—	1,417	—	—	—
Wyoming	4,994	4,990	—	4	—	—	—
Colorado	5,171	4,870	—	301	—	—	—
New Mexico	1,340	1,340	—	—	—	—	—
Arizona	7,366	3,125	431	—	3,810	—	—
Utah	9,904	9,770	—	134	—	—	—
Idaho	1,318	1,000	—	318	—	—	—
Nevada	2,500	2,500	—	—	—	—	—
Pacific							
Total	39,525	1,300	4,731	13,379	18,786	1176	153
Washington	16,198	—	—	10,782	5,416	—	—
Oregon	4,409	500	150	56	3,650	—	53
California	18,918	800	4,581	2,541	9,720	1176	100
West of Mississippi							
Totals	153,884.5	82,586	12,363.5	20,253	36,487	1176	1019
Percent	100.0	53.7	8.0	13.1	23.7	0.8	0.7

Table 5-29.

Electric power by fuel source for the year 2000 by comparison with 1975 (quads)

Region	Total	Coal	Liquid fuels	Natural gas	Nuclear energy	Hydro, geothermal, and solar energy
New England	1.8	0.1	0.5	0	1.1	0.1
Middle Atlantic	6.2	1.6	1.0	0	2.6	1.0
East North Central	8.2	4.8	0.3	0	3.1	0
West North Central	3.2	1.5	0.4	0.4	0.6	0.3
South Atlantic	7.3	4.9	0.9	0	1.1	0.4
East South Central	3.5	1.9	0.1	0	1.4	0.4
West South Central	6.5	2.0	0.3	1.6	2.5	0.1
Mountain	3.2	2.2	0.1	0.2	0.2	0.5
Pacific (including Hawaii and Alaska)	6.4	1.6	0.2	0	1.9	2.7
Hawaii and Alaska	0.3	–	–	–	–	–
U.S. total[a]	46.6	20.6	3.8	2.2	14.5	5.5
U.S. percent (1975)	100.0	44.5	15.1	15.6	15.7	9.0
U.S. percent (2000)	100.0	44.2	8.2	4.7	31.1	11.8

[a]Because of rounding, totals may not add.

Table 5-30.

Summary of total and sector energy inputs, 1975 and 2000
(quads)

Sector	Total	Coal	Oil	Gas	Electricity	Geothermal, solar, and other
1975						
Transportation	18.6		17.9	0.6	0.1	—
Residential-commercial	25.8	0.2	5.7	7.6	12.2	—
Industrial	25.9	3.8	5.5	8.5	8.0	—
Total	70.3	4.0	29.1	16.7	20.3	
2000						
Transportation	22.2	0	21.2	0.6	0.4	0
Residential-commercial	34.4	0.3	0.9	6.0	27.2	0
Industrial	44.4	10.8	3.4	9.3	19.0	2.0
Total	101.0	11.1	25.5	15.9	46.6	2.0

Table 5-31.

Estimated regional energy consumption in the year 2000[a]
(quads)

Region	Total	Household	Commercial	Industry	Transportation
U.S. total	101.0	21.8	12.6	44.4	22.2
New England	4.4	1.6	0.9	1.1	0.8
Middle Atlantic	14.4	4.1	2.4	5.5	2.6
East North Central	19.9	4.8	3.0	9.2	3.1
West North Central	6.8	1.9	1.0	2.1	1.8
South Atlantic	13.1	2.8	1.6	5.3	3.4
East South Central	7.5	1.1	0.5	4.3	1.6
West South Central	17.0	2.0	1.1	10.2	3.7
Mountain	5.6	1.3	0.7	2.2	1.4
Pacific	11.9	2.2	1.3	4.6	3.9

[a]As a result of rounding, figures may not add to totals.

REFERENCES

1. Kenneth W. Costello, *Regional Demand Modeling,* Argonne National Laboratory, Argonne, Illinois, August 1976.

2. United States Department of the Interior, Bureau of Mines Information Circular, *Fuels and Energy Data: United States by States and Census Divisions,* U.S. Government Printing Office, Washington, D.C., 1976.

3. Federal Power Commission, news release, October 20, 1976.

4. E. L. Allen *et al., U.S. Energy and Economic Growth, 1975-2010,* Publication ORAU/IEA-76-7, Institute for Energy Analysis, Oak Ridge Associated Universities, Oak Ridge, Tennessee, 1976.

5. *Fossil Energy Study,* Institute for Energy Analysis, Oak Ridge Associated Universities, Oak Ridge, Tennessee, 1977, especially Chapter 2, p. 2.

6. In 1975, the average household size was 2.92. With 72×10^6 households and a population of 213×10^6, this implies that 98 percent of the population belonged to households.

7. U.S. Department of Transportation, Federal Highway Administration, *Highway Statistics, 1975,* Section 1, Vehicles, Drivers, and Fuels, Publication FHWA-HP-HS-75-OIR, Washington, D.C., 1975.

8. Data were derived from unpublished tables by the Federal Highway Administration.

9. Bureau of Mines, U.S. Department of the Interior, *Sales of Fuel Oil and Kerosene in 1975,* Mineral Industry Surveys, Washington, D.C., September 17, 1976, p. 11.

10. Independent Petroleum Association of American, *The Oil Producing Industry in Your State,* Washington, D.C., 1976, p. 99.

11. The Conference Board in cooperation with the National Science Foundation, *Energy Consumption in Manufacturing,* Ballinger, Cambridge, Massachusetts, 1974. Other sources included: Charles Berg, "Conservation in Industry," *Science,* vol. 184, pp. 267-270 (1974); "A Technical Basis for Energy Conservation," *Technology Review,* vol. 76, pp. 15-24, February 1974; Hittman Associates, Inc., *Environmental Impacts, Efficiency, and Cost of Energy Supply and End Use,* Columbia, Maryland, 1974, vol. 1; Stanford Research Institute, *Patterns of Energy Consumption in the United States,* Stanford, California, 1972.

12. Office of Energy Systems, Federal Power Commission, *State Projections of Industrial Fuel Needs,* Washington, D.C., August 1976.

13. Feedstocks, which add another 50 percent to fuel needs in the petrochemical industry, are not included here because of lack of information. We have added these separately in a later section.

14. Bureau of Economic Analysis, *Area Economic Projections, 1990,* Washington, D.C., 1974.

15. According to the study on petroleum refining in *Energy Consumption in Manufacturing* [see (11)], the annual rate of decline between 1962 and 1971 was 2.0 percent in net energy.

16. Lulie H. Crump, *Fuel and Energy Data: United States by States and Regions,* U.S. Bureau of Mines Publication IC 8739, Washington, D.C., 1974.

17. Federal Power Commission, *FPC News* (week ending February 20, 1976), U.S. Government Printing Office, Washington, D.C., 1976, p. 22.

18. *Projects to Expand Fuel Sources in Western States,* U.S. Bureau of Mines Publication IC 8719, Washington, D.C., 1976; *Projects to Expand Fuel Sources in Eastern States,* U.S. Bureau of Mines Publication IC 8725, Washington, D.C., 1976.

6

**FUTURE PRICES AND
ELASTICITIES**

Introduction

Although the IEA approach to energy-demand modeling is basically non-econometric, there are imbedded in the approach a set of implicit price-quantity relationships. These relationships between prices and energy demands are not based on the unwavering assumption that the future will be identical to the past. But it is still a useful exercise to establish whether or not there is a reasonable historical precedent for the conservation scenarios developed in this report.

To make this comparison, it is necessary to derive imputed elasticities from the underlying price-quantity relationships, which were developed with non-econometric tools. The price elasticity of demand, which we shall denote as ϵ_{ij}, gives the percentage change in the quantity of commodity i demanded when the price of commodity j changes by 1 percent. A related quantity is the income elasticity of demand, which we shall write as η_i. This term gives the percentage change in the quantity of commodity i demanded when income changes by 1 percent (holding all prices constant).

Elasticity Basics

When we investigate the effect of a commodity's own price on its demand, i and j are the same commodity, and ϵ_{ii} is called simply the "own price elasticity." When i and j are different commodities, we have what are known as "cross elasticities."

Since the concept of elasticity is stated in terms of percentage changes in demand and percentage changes in price, the measure is independent of any units of measure and currency prices. Thus it is possible to compare directly energy studies conducted in West Germany and the United States, despite the fact that the energy data in the German study are in joules whereas in the American study the units are Btu's and prices in Germany are in deutsche marks and in America are in dollars. This remarkably convenient property makes elasticities the most useful measure available of demand responsiveness to price changes.

Own elasticities are generally (although not always) the most interesting magnitudes to measure, and several bench marks exist against which to judge price responsiveness. If the elasticity is zero, then demand is said to be perfectly inelastic; that is, price has no effect on demand. If the elasticity is -1, then a 1 percent increase in price will cause demand to decrease by 1 percent; in this case, demand is said to be unitary elastic. If the price elasticity of demand is between zero and -1, then demand is said to be inelastic, and an elasticity larger than one in absolute value is termed elastic. Elasticities can be associated with demand curves, and the graphs of these demand curves and matching elasticities are depicted in Figure 6-1.

In this report only negative own elasticities have been considered; that is, price increases have a negative impact on the quantity demanded. Economic theory suggests that, as prices rise, the quantity demanded decreases, other things being constant, so that, if ϵ_{ii} is positive, the response can be said to be perverse, since higher prices would induce consumers to purchase more. Such a consequence is theoretically possible but highly unlikely, and there exists no generally accepted empirical exception to the rule.

Although own elasticities should always be negative, cross elasticities can take on any values. In fact, if a cross elasticity is negative, then the two commodities can be said to be substitutes. For example, electricity and natural gas are substitutes if the demand for natural gas increases as the price of electricity increases. If two commodities are complements, then ϵ_{ij} is positive. For example, as gasoline prices increase, the demand for automobiles decreases.

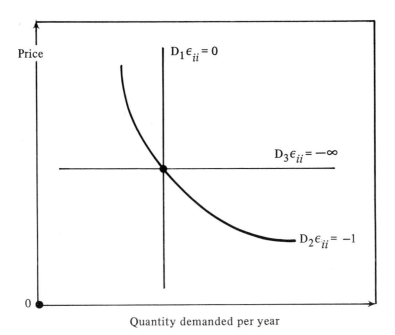

Figure 6-1.
Own price elasticity and demand curves designated by D_1, D_2, and D_3

We may use the concept of elasticity to deduce the relationship between changes in price and income and the resulting changes in demand. This relationship may be expressed as the following system of equations:

$$\hat{Q}_c = \epsilon_{cc} \hat{P}_c + \epsilon_{c\ell} \hat{P}_\ell + \epsilon_{cg} \hat{P}_g + \epsilon_{ce} \hat{P}_e + \eta_c \hat{Y} \tag{1}$$
$$\hat{Q}_\ell = \epsilon_{\ell c} \hat{P}_c + \epsilon_{\ell\ell} \hat{P}_\ell + \epsilon_{\ell g} \hat{P}_g + \epsilon_{\ell e} \hat{P}_e + \eta_\ell \hat{Y} \tag{2}$$
$$\hat{Q}_g = \epsilon_{gc} \hat{P}_c + \epsilon_{g\ell} \hat{P}_\ell + \epsilon_{gg} \hat{P}_g + \epsilon_{ge} \hat{P}_e + \eta_g \hat{Y} \tag{3}$$
$$\hat{Q}_e = \epsilon_{ec} \hat{P}_c + \epsilon_{e\ell} \hat{P}_\ell + \epsilon_{eg} \hat{P}_g + \epsilon_{ee} \hat{P}_e + \eta_e \hat{Y} \tag{4}$$

where $\hat{Q}_i$ is the percentage change in the quantity of commodity i demanded, with c denoting coal, ℓ denoting liquid fuels such as petroleum products, g denoting natural gas, and e denoting electricity; $\hat{P}_i$ is the percentage change in the price of commodity i; and $\hat{Y}$ is the percentage change in income. This expression can also be written somewhat more simply as

$$\hat{Q}_i = \sum_j \epsilon_{ij} \hat{P}_j + \eta_i \hat{Y} \tag{5}$$

We can solve each of these equations for any given elasticity by simple algebraic manipulation of Eq. 5. Solving for own elasticities gives

$$
\epsilon_{ii} = \frac{\hat{Q}_i}{\hat{P}_i} - \sum_{\substack{j \\ j \neq i}} \epsilon_{ij} \frac{\hat{P}_j}{\hat{P}_i} - \eta_i \frac{\hat{Y}}{\hat{P}_i}
\tag{6}
$$

The Scenario Tests

We can use this framework to analyze the underlying price-demand relationship in the energy sector. Two cases have been considered, with net energy consumption levels of about 101 and 126 quads. A "business as usual" reference case is also given for comparison purposes.

The reference case differs from the 101- and 126-quad cases in two assumptions. First, it allows for the population, gross national product (GNP), and energy impacts of a continuing influx of illegal immigrants. Second, it adjusts consumption to accommodate larger availabilities and lower prices for petroleum and natural gas. Total energy requirements for the reference case are 150 quads. This represents an average annual rate of increase of 2.15 percent over the period 1975 to 2000, which on a per capita basis would be well above the average for the period 1940 through 1975.

Energy prices for the year 2000 are given in Table 6-1, and their derivation

Table 6-1.

Energy price indexes[a] in the year 2000
(relative to 1975 in constant dollars)

Fuels	Case (quads)		
	101	126	Reference
Coal	1.65	1.65	1.00
Oil	2.40	2.40	1.20
Gas	10.00	10.00	5.00
Electricity	1.65	1.65	1.00

[a]The 1975 average prices were as follows: coal, $17.50 per ton, delivered to utilities; oil, $10.40 per barrel, composite cost to refiners; natural gas, $0.43 per thousand cubic feet at the wellhead; electricity, 27 mills per kilowatt-hour to consumer.

is given in Appendix C. Table 6-1 gives energy prices in index form. Each fuel's index is adjusted so that its price equaled unity in 1975. Prices for the reference case are much lower than for the other two cases, and the additional energy demanded can be attributed to lower energy prices. There is no price differential between the 126-quad case and the 101-quad case, and the demand difference between the two depends on three factors: the rates of growth of both population and labor productivity are higher in the 126-quad case than in the 101-quad case and the responsiveness of buyers to higher energy prices is assumed to be greater.

In Table 6-2 we list the population and GNP assumptions for the year 2000. Table 6-3 summarizes energy demands for the three cases and four energy sources. Although the two IEA cases yield 101 and 126 quads, respectively, totaling the energy demands yields numbers that are seemingly too large. For example, for the 101-quad scenario, totaling coal, liquid fuels, gas, and electricity demands yields 126 quads. This does not mean that there are 126 quads of energy available for the use of society. There are only 101 quads, because 25 quads of coal, oil, and natural gas were converted to electricity. Even though totaling the sales of all energy carriers gives a socially irrelevant number, the individual outputs of coal, oil, gas, and electricity are indicative of the total demands for these inputs and are the appropriate numbers to use in computing elasticities. Coal, gas, and oil are all used to produce electricity.

Table 6-2.

Population and GNP assumptions for the year 2000
(population x 10^6, GNP x 10^9 $1972)

| Parameter | Case (quads) | | |
	101[a]	126[a]	Reference[b]
Population	246	254	260
GNP	$2620	$2648	$2822

[a]For the 101- and 126-quad cases, GNP growth is 3.8 percent, 1975-1985, and 2.8 percent, 1985-2000.

[b] For the reference scenario, GNP growth is 4.0 percent, 1975-1985, and 3.8 percent, 1985-2000.

Table 6-3.

Gross energy demands for the three cases and four energy carriers for the year 2000

Energy carrier, direct use	Case (quads)[a]		
	101	126	Reference
Coal	31.6	39.3	38.1
Liquid fuels	29.2	38.4	46.4
Gas	18.1	18.1	31.6
Electricity	46.6	61.8	70.8
Other	2.0	2.0	2.0

[a]Columns do not total 101 and 126 quads, respectively, because of the shift of a part of the coal, liquid fuel, and gas use to electricity.

Adding the number of Btu's of coal going into the generation of electricity and then counting the energy value of the electricity leads to double counting of energy in an aggregate sense. The problem at hand requires that we focus on the total demands for each energy carrier. It is, in fact, important to take into account the interaction of coal and electrical energy demands.

Since we can calculate all of the $\hat{Q}_i$'s, $\hat{P}_i$'s, and $\hat{Y}$ in Eqs. 1 through 4, we are well on our way to calculating the own elasticities. The only problem is that we do not know any of the cross elasticities. From looking at this set of equations, one can see that the set of own and cross elasticities is not unique to the system. There are 20 unknowns but only four equations. If values for any 16 of the variables are known, then the own elasticities can be calculated. The problem becomes one of finding appropriate values for the cross elasticities.

In reality, cross and own elasticities will have true empirical values. But even a short review of the econometric literature is enough to convince anyone that there is little agreement concerning the values of own price elasticities, to say nothing of cross elasticity values. As a consequence, we have made the following simplifying assumptions concerning income and cross elasticities. Income elasticities, η, are all unity. Cross elasticities are all assumed to be equal to zero. Following this procedure, we are able to calculate "gross" own elasticities, since Eq. 6 now reduces to

$$\epsilon_{ii} = \frac{\hat{Q}_i}{\hat{P}_i} - \eta_i \frac{\hat{Y}}{\hat{P}_i} \tag{7}$$

Results of this calculation are given in Table 6-4.

A glance at Eq. 6 shows that, if all commodities are substitutes for one another, that is, $\epsilon_{ij} > 0$, where $i \neq j$, then gross elasticities are underestimates (their absolute value is too low). To the extent that energy carriers are complements, we have overestimated the absolute values of our elasticities.

The assumption that cross elasticities between oil and other products are equal to zero would seem justified when we recognize that almost all of the petroleum in 2000 in our scenarios is allocated to the transportation sector, where little substitution seems possible. The use of coal, as we mentioned earlier, is very closely tied to the generation of electricity (about two-thirds of all the coal produced is assumed to go into the generation of electricity). Coal and electricity are quite likely to be complements rather than substitutes. But coal, oil, and gas will be substitutes in the generation of industrial process heat. As a consequence, lower prices for liquid fuels and natural gas drew customers away from coal and lowered total demand in spite of coal's lower price. The unaccounted-for cross effects are responsible for the positive own elasticity for coal in the 126-quad case.

The natural gas sector presents a peculiar problem, since its output is assumed to be extremely supply-constrained in the 101- and 126-quad sce-

Table 6-4.

Gross own elasticities of demand
(calculated with respect to the
reference case)

Elasticity	Case (quads)	
	101	126
ϵ_{cc}	−0.23	+0.11
$\epsilon_{\ell\ell}$	−0.35	−0.13
ϵ_{gg}	−0.41	−0.39
ϵ_{ee}	−0.50	−0.13

narios. The extremely low relative price of natural gas in the reference scenario reflects the assumption of new discoveries.

Conclusion

We come finally to an evaluation of our elasticities in relation to the historical record. To do this we compare the elasticities in Table 6-4 to those obtained by econometric researchers. A survey of the econometric literature can be found in our IEA study (1). The electricity and natural gas demand elasticities in Table 6-4 are well within the range of econometric findings for long-run elasticities and consequently seem conservative estimates. The few studies that have been conducted on the demand for coal indicate that the long-run elasticity probably lies somewhere between -0.5 and -1.5. Our gross elasticity, although somewhat less, does not disprove these figures. There have been numerous studies conducted on the elasticity of demand for petroleum products (2), and long-run elasticities vary greatly. Our gross elasticities are well within the range of econometric findings.

REFERENCES

1. J. Edmonds, "A Guide to Price Elasticities of Demand for Energy: Studies and Methodologies," Institute for Energy Analysis, Oak Ridge Associated Universities, Oak Ridge, Tennessee, August 1978.

2. R. G. McGillioray, "Gasoline Use by Automobiles," Working Paper 1216-2, Urban Institute, Washington, D.C., August 1974; J. Ramsey, R. Rasche, B. Allen, "An Analysis of the Private and Commercial Demand for Gasoline," *Review of Economics and Statistics,* vol. 57, No. 4, November 1975, pp. 502-507; J. Sweeney, "Passenger Car Use of Gasoline: An Analysis of Policy Options," Federal Energy Administration, Washington, D.C., February 1975.

APPENDIX A

ILLEGAL IMMIGRATION AND FUTURE ECONOMIC GROWTH

Introduction

Our projections of future economic growth have relied heavily on two primary contributing factors: (i) the increasing size of the labor force and (ii) the growth of factor productivity. The latter is a dollar measure of the change in the annual (or hourly) output per worker. It is determined by an amalgam of factors, including increases in the level of education and capital investment and those technological advances that lead to greater efficiency. The size of the labor force depends upon the participation rate (the ratio of persons 16 years and older working or actively seeking employment to the total population 16 years and older).

The rate of population growth in the United States clearly is a primary consequence of the fertility of American women, as well as of the rate of net immigration. The U.S. fertility rate (FR) has now fallen well below the level needed to sustain the population at the replacement level in the long run without substantial immigration. Projections of the U.S. population to the year 2000 based on various assumptions with respect to future fertility have been made by the Census Bureau. All of these projections include an annual net increment of 400,000 immigrants (1). This level of immigration and its composition is largely a result of Public Law 89-236, passed in 1965, which abolished the former national-origins system of immigration controls. It established limits for the Western Hemisphere (120,000) and the Eastern

Hemisphere (170,000); immediate relatives of U.S. citizens are allowed to enter outside the quota. Actual annual immigration, as recorded by the Census Bureau, averaged 321,000 in the years 1972 through 1976, excluding the emergency entry of 130,000 Vietnamese refugees in 1975 (2). Some 14.5 percent of the legal immigrants who entered the United States in the decade 1960 through 1970 are estimated to have left by 1970, but the Census Bureau does not subtract out all returnees in its population projections (3). Uncertainties about the size of legal flows, therefore, are statistically significant.

Moreover, the flow of persons arriving illegally in the United States is only very slightly reflected in current estimates or future projections of population made by the Census Bureau, since there is no specific allowance in these estimates for illegal entrants. However, illegal immigration has grown to such a size that, even with returnees subtracted out, it is now believed by many experts to exceed the legal immigration (4). If these estimates are correct, then perhaps 40 percent of the current population growth in the United States is due to immigrants, legal and illegal.

Illegal Immigrants Currently in the United States

There are widely disparate estimates of the number of illegal aliens now residing in the United States, and of the number of these who are employed. In a major study the Immigration and Naturalization Service (INS) of the Department of Justice polled its districts to obtain estimates of national totals (5). These district estimates indicated that there are in the United States between 5.5×10^6 and 6.0×10^6 illegal aliens, of whom some 3.7×10^6, or about 63 percent, are believed to be employed. Other estimates by private researchers (6) ranged up to 8×10^6. The most recent federal estimate is much lower, between 3×10^6 and 5×10^6 (4).

It is not surprising that estimates of the total number of illegal aliens (often referred to in official literature as undocumented aliens) vary widely. A spokesman for the Census Bureau is quoted as believing that the 1970 census omitted 5.3×10^6 people who were residing in the United States at the time of the count (7). The undercount was thought to be a result of the omission of blacks and was not believed to be primarily a reflection of the number of illegal aliens. Hence, since illegal aliens attempt to avoid identification, their numbers are not subject to rigid statistical verification.

The number of illegal immigrants entering the United States each year also is subject to a wide range of estimates. The estimates, which are based on the

annual numbers of illegal aliens apprehended, the volume of tips received by
the INS, and direct evidence, are not simply judgments. The bulk of those
apprehended are taken at the border and therefore do not enter the U.S.
labor market. Those apprehended also include an unknown number of "re-
peaters." A study in fiscal year 1974, based on sensor data, of the number of
apprehensions (A) at the Mexican border and estimates of those who entered
unapprehended (gotaways or G) showed a G/A ratio of 1.8:1 (8). However,
this ratio almost certainly is a substantial overestimate, since (as we learned in
Vietnam) many things activate sensors in addition to people engaged in illegitimate activity. If this ratio is reasonably correct, then the annual gross
inflow is very high across the Mexican border. Historically, Mexico has been
the major source of unlawful entrants into the United States and is expected
to remain so (see Table A-1). Many other developing countries also contribute
to the flow. A good number of these countries are expanding their economies
at a rapid pace, but birth rates are usually very high and employment opportunities in these countries are grossly inadequate.

Nearly half of Mexico's population of 62×10^6 is under 15 years of age.
Reportedly, 25 to 50 percent of those of labor force age are either unemployed or underemployed (9). The estimated growth of Mexicans of labor
force age between 1970 and 1985 is 75 percent (9); if the Mexican economy
were to absorb all these potential workers, a very high and almost certainly
unreachable rate of economic expansion would be required. The annual population growth in Mexico in 1976 was 3.5 percent, one of the highest in the
world, compared to 0.8 percent in the United States (9). Former INS Com-
missioner Leonard Chapman has estimated that, in addition to border
crossers, up to 10 percent of the 6×10^6 annual visitors who enter the United
States legally cannot be accounted for in departure data.

Warren and Peck, who studied emigration from the United States in the
decade of the 1960s (3), say that 25 percent of the legal Mexican entrants
during this decade voluntarily returned to Mexico. Adjustment pressures on
illegal entrants must be much greater than on legal ones; we therefore assume
that the number of illegal immigrants returning to Mexico is higher than the
number of legal immigrants. Because of the geographic proximity of their
country (most enter the labor market in the southwest region of the United
States), Mexicans can return home more easily than illegal immigrants from
virtually any other country except Canada. The majority of aliens from countries other than Mexico who are here illegally generally settle permanently.

Table A-1.
Major source countries of illegal aliens[a]

Country	Population estimate, mid-1976 (x 10⁶)	Annual rate of population growth (percent)	Number of years to double population	Population projection to 2000 (x 10⁶)	Population under 15 years (percent)	Urban population (percent)	Per capita gross national product (U.S.)
United States (reference)	215.3	0.8	87	262.5	27	74	6640
Mexico	62.3	3.5	20	134.4	46	61	1000
Dominican Republic	4.8	3.0	23	10.8	48	40	590
Haiti	4.6	1.6	43	7.1	41	20	140
Jamaica	2.1	1.9	36	2.8	46	37	1140
Guatemala	5.7	2.8	25	11.1	44	34	570
Colombia	23.0	3.2	22	44.3	46	64	510
Peru	16.0	2.9	24	30.9	44	60	710
Ecuador	6.9	3.2	22	14.0	47	39	460
Philippines	44.0	3.0	23	86.3	43	32	310
Korea	34.8	2.0	35	52.3	40	41	470
Thailand	43.3	2.5	28	86.0	45	13	300
Greece	9.0	0.4	173	9.7	25	53	1970
India	620.7	2.0	35	1051.4	40	20	130
Iran	34.1	3.0	23	67.0	47	43	1060
Nigeria	64.7	2.7	26	135.1	45	16	240

[a]Population data are from the Population Reference Bureau, Washington, D.C. (reproduced in Domestic Council Committee on Illegal Aliens, *Preliminary Report*, Washington, D.C., December 1976, p. 46).

The latest official estimate of the annual net inflow of immigrants into the
United States is 500,000 (4). As will be shown later, this is a significant number in terms of economic growth. Of course, our primary concern is about
the future. But a few key points need to be made about the recent past. The
illegal alien has been recognized as an increasing economic and social problem
for about 10 years; in this time period, the number of illegal aliens apprehended has increased from about 100,000 to 800,000 a year. Most of the
illegal immigrants still in the United States, therefore, probably arrived after
1965. Their contribution to past domestic economic growth has been very
largely accounted for in Department of Commerce estimates of the annual
increments to the gross national product (GNP), since these calculations of
the output of goods and services are estimated totals for the nation. Many
illegal aliens, particularly those who are assimilated, are probably counted in
the official labor force data, particularly those that have Social Security
cards. Because the Census Bureau population estimates do not include many
illegal aliens, population estimates of these people are understated. But we
do not know by how much, since some illegal aliens probably are included in
the Census counts. This means that per capita statistics of productivity increases from 1965 through 1975, and of energy consumption, have been very
modestly overstated.

Proposed Legislation

Before turning to future estimates, let us consider the proposed Carter Ad-
ministration legal actions. Formally announced in a message to Congress on
August 4, 1977, the Carter program includes both administrative directives
and new legislation. The legislation, under consideration by the Judiciary
Committees of the Congress, is called the Alien Adjustment and Employment
Act of 1977. The major features of the President's program are as follows:

1. Make unlawful the hiring of undocumented aliens, with enforcement by
the Justice Department against those employers who engage in a "pattern or
practice" of such hiring. Penalties would be civil—injunctions and fines of
$1000 per undocumented alien hired. Criminal penalties could be imposed
by the courts against employers violating injunctions. Moreover, employers
and others receiving compensation for knowingly assisting an undocumented
alien to obtain or retain a job would also be subject to criminal penalties.

2. Increase significantly the enforcement of the Fair Labor Standards Act

(which mandates payment of the minimum wage) and the Federal Farm Labor Contractor Registration Act (which prohibits the hiring of illegals for farm work), targeted to areas where heavy hirings of undocumented aliens occur.

3. Adjust the immigration status of undocumented aliens who have resided in the United States continuously from before January 1, 1970, to the present and who apply to the Immigration and Naturalization Service for permanent resident alien status; create a new immigration category of temporary resident alien for undocumented aliens who have resided in the United States continuously prior to January 1, 1977; make no status change and enforce the immigration law against those undocumented aliens entering the United States after January 1, 1977.

4. Substantially increase resources available to control the southern border, and other entry points of the United States, in order to prevent further illegal immigration and control alien smuggling rings.

5. Promote continued cooperation with the governments of those nations which are major sources of undocumented aliens, in an effort to improve their economies and their employment opportunities.

Future Trends

The extent of the future flow of illegal aliens probably depends most importantly on U.S. immigration policy. Other industrialized countries control the entry of aliens by requiring work and residence permits, registration with the police, and other devices to discourage illegal entry. Hence, an increase in the number of illegal entrants in the United States is probably not inevitable. The size of the current flow could almost certainly be sharply cut back if the legislative decision were made to do so. Given the lack of Congressional action on illegal entrants when U.S. unemployment increased from 4×10^6 in 1970 to nearly 8×10^6 in 1975, the prospects for action in the near future are not bright. This is not to say that forceful Congressional action will not be taken at some point in the future if illegal flows become a threat to domestic social and economic institutions. Therefore, we consider below a range of numbers for illegal entrants which represents possible future flows in this century. At the bottom of the range, a tight control program, adequately financed and administered as effectively as controls exercised by some industrialized Western European countries, could probably reduce the illegal flow to a fraction of the legal one and thereby eliminate the problem. We label this possibility case A. Under this case, illegal aliens would become again, as they have often been in the past, a nonproblem. The United States has always had some illegal aliens crossing its borders and entering the labor market. However, these flows have usually not been of such a size as to threaten economic problems or disruption of organized labor markets. The other cases considered, in terms of the annual entry of illegal aliens, are

	Case B	Case C	Case D
Gross number of illegal aliens entering the United States	500,000	1,000,000	1,500,000
Returnees (35%)	175,000	350,000	525,000
Net annual inflow	325,000	650,000	975,000
Cumulative population addition by the year 2000	10,914,000	21,829,000	32,743,000
Percent	4.5	8.9	13.4

This range of possible annual flows probably brackets the size of the illegal flows the United States is likely to experience over the next several decades. Some experts believe that the high levels (cases C and D) are unlikely to be sustained over the long term (10). Their reasoning is that flows of this magnitude would stimulate dramatic changes in laws governing immigration, since this number of illegal immigrants would be much more visible and the problems they would create would be large enough to require special relief and employment measures.

The comparison base for the cumulative population addition by the year 2000 used above is with the future population estimates of the Census Bureau Series III projection, which is computed on the basis of a completed FR of 1.7 children per female. However, for females of illegal resident families, we have used an FR of 2.1 children per female, largely for computational convenience, so that the Census Bureau Series II and II-X projections could be used. Contrary to the stereotype, the fertility of immigrant women is about the same as that of native-born women. The Census Bureau Series II-X projection to the year 2000 assumes no immigration, legal or otherwise (11). The cumulative impact of an annual net flow of 400,000 immigrants on population growth through 2000 can be derived by subtracting the Census Bureau

Series II-X from their Series II projections. Both of these projections use an FR of 2.1 children per female. In the absence of any reliable information on the age-sex distribution of illegal aliens, the age-sex distribution of legal immigrants has been used even though there may be significant differences. Over time, however, the age-sex distribution of illegal immigrant families probably approaches that of legal ones.

Labor Force Additions

The percentage of the total illegal population 16 years and over was estimated to be the same over time as for the population as a whole and therefore to range from 73.5 percent to 76.1 percent in individual future years. The resulting estimates of those 16 years and older were then adjusted for national labor force participation rates, which were studied independently. Participation is expected to rise from the current rate of 61 percent to 63.5 percent by the year 2000. The estimates are shown in Table A-2.

Impact on GNP

Before translating increased labor force numbers into an estimate of the impact on the total output (GNP), we need to make a further adjustment. Most apprehended illegal aliens are relatively uneducated and lack English language

Table A-2.

Illegal aliens: cumulative additions to the labor force[a]
(in thousands of people)

Years	Case B		Case C		Case D	
1976-1980	789	(742)	1578	(1484)	2,367	(2,226)
1981-1985	1711	(1643)	3422	(3286)	5,133	(4,929)
1986-1990	2709	(2601)	5418	(4828)	8,127	(7,803)
1991-1995	3790	(3638)	7580	(7276)	11,370	(10,914)
1996-1999	4996	(4796)	9992	(9592)	14,899	(14,388)

[a]Figures in parentheses are employed workers, assuming 6 percent unemployment in 1976-1980 and 4 percent thereafter. These numbers are not adjusted to U.S. equivalents. We have made no attempt to compute a seasonal employment factor; this correction is in part ground into the calculated difference between gross and net numbers of aliens.

skills. Apprehended illegal aliens from Mexico, for example, had an average of 4.9 years of education, compared with 12.3 years for the U.S. population over 25 years of age. Further, there is reported to be a downward occupational shift for the more educated illegal aliens from white-collar to service jobs after they enter the United States (12).

The North and Houstoun study, based on a sample of 793 illegal aliens working in 1975 and apprehended, showed them receiving an average wage of $2.71 an hour (12). The average work week was 44.5 hours for the sample group, for an annual income of about $6000. Although probably below wages paid U.S. workers for comparable jobs, the differential is to some extent a reflection of lower productivity. Wage and salary data and counterpart income of self-employed persons in 1975 came to an average of $11,850 per employed person. Hence, wages to illegal workers were about half the U.S. average. The U.S. average reflects the far higher contribution to output of those employers and employees with supervisory and professional skills. Our employed illegal alien GNP/worker average, then, is 50 percent of the national GNP/worker average. The 50 percent wage factor, of course, is not a measure of jobs held by aliens and therefore is not a valid measure of their impact on the domestic job market, which would be greater.

Recapitulation of Methodology

To make the method of computation somewhat clearer, we recap it below. First of all, if we subtract the Census Bureau Series II-X population projection from their standard Series II, the residual is an estimate of the cumulative population additions due to immigration (400,000 a year). This estimate includes the long-term consequences of a fertility rate of 2.1 children per female in the families of illegal aliens. We then assume that the population impact of illegal entrants will be the same as for legal ones, and we adjust only for differences in the relative size of the flow in our three calculated cases.

Our Case B estimates a net annual inflow of 325,000 illegal immigrants per year (gross entry, 500,000), or 81.25 percent of 400,000. The legal immigrant series is then adjusted for each year, 1975 to 2000, by a multiplier of 0.8125. Similarly, the Case C series, 650,000 net annual inflow (gross entry, 1,000,000), is adjusted by a multiplier of 1.625 and the Case D, 975,000 net inflow (gross entry, 1,500,000), by a multiplier of 2.44. Employed labor force estimates, by year, in each case are calculated from the following four factors:

1. Census Bureau data on the ratio of persons 16 years and older to the total population for each year, 1975 to 2000, are the basis of the year-by-year calculations of illegal aliens of labor force age.

2. The estimated proportion of those in the working age pool who will elect to participate in the labor force has been derived from independent analyses. It ranges from 61 percent in 1975 to 63.5 percent in 2000.

3. The rate of unemployment has been arbitrarily set at 6 percent through 1980 and at 4 percent thereafter.

4. Based on estimates of wage differentials, the number of illegal aliens estimated to be in the labor force is adjusted to U.S. equivalents. (On the average, one illegal alien equals 0.5 U.S. worker in terms of his contribution to the GNP.)

The final estimates are subject to an unknown margin of error attributable to the arbitrary assumptions that were made. Among the assumptions, the following are probably most subject to error:

1. The assumption that the age-sex ratio of illegal entrants will be approximately the same as that for legal migrants.

2. The assumed FR of 2.1 children per female for families of illegal entrants; this figure is well above the value of 1.7 children per female we have used for the domestic population.

The total additions to the labor force by illegal aliens based on the above methodology are shown in Table A-3.

Impact of Illegal Aliens on Rates of Economic Growth

By comparing the annual labor force additions in the cases shown in Table A-3 with the annual additions that would take place in the absence of a flow of illegal aliens, we can estimate the change in the rate of economic growth that would result from flows of various sizes. The population projection that we have considered most likely is the Census Bureau Series III. The long-run implications of declining FR on the growth of the labor force since 1957 (3.8 children per female in 1957 to 1.8 in 1975) and of the moderate continuing decline to a rate of 1.7 in themselves will cause sharp reductions in the future rate of economic growth. For example, the current annual rate of labor force growth (1975 to 1980) is 1.8 percent; by 2000 it will fall to 0.3 percent annually, assuming 400,000 immigrants a year and an FR of 1.7 children per female. (These estimates do not imply greater than normal levels of unemployment in the United States or a decline in the rate of growth of

Table A-3.

Total employed labor force, with and without illegal aliens[a]
(thousands of people)

Years	Estimated total labor force excluding aliens	Illegal aliens added (in U.S. equivalents)					
		Case B		Case C		Case D	
1976-1980	93,934	391	(0.42)	742	(0.79)	1114	(1.19)
1981-1985	102,848	821	(0.80)	1642	(1.60)	2464	(2.40)
1986-1990	109,244	1300	(1.19)	2600	(2.40)	3900	(3.57)
1991-1995	112,914	1819	(1.61)	3637	(3.22)	5456	(4.83)
1996-2000	116,291	2398	(2.06)	4795	(4.12)	6442	(5.54)

[a]Data are annual averages for the 5-year periods shown. Figures in parentheses are annual percentage increases in each 5-year period. Labor force additions include children of illegal aliens reaching 16 years of age.

per capita incomes.) A marked reduction in the rate of economic growth as the U.S. population approaches a maximum size (for an FR of 1.7 children per female, the population peaks in 2020 at 253×10^6) seems most probable, even if productivity moves back to postwar highs. The flow of illegal entrants, if large enough and sustained long enough, could modify this forecast, with consequent expansionary effects on both the GNP and energy demand.

Tables A-4 and A-5 are a first approximation of the differences in future economic growth, assuming that one-half of the GNP growth is directly attributable to the expansion of the labor force. This percentage has been used because 53.5 percent of the change in the GNP between 1946 and 1975 was accounted for by the change in wages and salaries during the period. The GNP growth calculations in the column labeled "annual GNP increase excluding illegal aliens" are taken from an earlier IEA study (13). Table A-4 shows that, if the illegal flow began in 1975 and amounted to a net influx of 650,000 persons or more, the labor force would increase by over 20 percent and the impact on economic growth (and hence on energy demand) would be significant by 1990.

If the annual net inflow were about 975,000 persons, then it would be suf-

Table A-4.

Annual additions to the employed labor force with and without illegal aliens

Years	Annual increase excluding illegal aliens (thousands)	Annual increase by illegal aliens					
		Case B		Case C		Case D	
		Thousands	%	Thousands	%	Thousands	%
1979-1980	1987	80	4.0	160	8.1	240	12.1
1984-1985	1527	90	5.9	180	12.0	270	19.7
1989-1990	850	96	11.3	192	22.5	290	34.1
1994-1995	606	110	18.2	214	35.3	321	52.9
1999-2000	739	119	16.1	239	32.2	357	44.2

Table A-5.

Annual estimated rates of GNP growth with and without illegal aliens

Years	Annual GNP increase excluding illegal aliens	Annual GNP increase with illegal aliens		
		Case B	Case C	Case D
1979-1980	3.8	3.9	4.1	4.3
1984-1985	3.2	3.4	3.6	3.8
1989-1990	2.7	3.0	3.3	3.6
1994-1995	2.5	3.0	3.4	3.8
1999-2000	2.5	2.9	3.3	3.6

ficient to sustain long-run economic growth at or above a rate of 3.6 percent annually, which would be about equal to the postwar growth trend in the United States. Under these conditions, the dampening down in energy demand would be more a consequence of energy price elasticities than of a decline in the rate of economic growth. At the other end of the range, an annual illegal flow of 200,000 to 300,000 would mean very little in the way of additional GNP growth. This rate of illegal flow has almost certainly already been included in the official economic growth calculations for 1965 through 1975. As long as the number of illegal aliens does not increase over the value in the past, we would not be inclined to adjust growth forecasts upward, since to do so would lead to double counting.

Finally, the productivity of the portion of the labor force attributable to illegal aliens will, over time, probably improve from the 50 percent of the average U.S. employed person that we have used in our calculations. The children of illegal aliens who complete U.S. secondary education programs and thereby acquire English language and technical skills thus become an important segment of the labor force from the standpoint of economic growth. These illegal aliens (and American-born children of illegal parents, who are not classified as illegal aliens) will probably approach and equal the GNP contribution of the average U.S. citizen with time. Some of this improvement is undoubtedly going on now (14). In summary, if illegal immigration approaches that specified in the high-entry case (Case D), or close to 10^6 net illegal immigrants per year, it could push the annual U.S. economic growth rate in the 1990s to or above 3.6 percent (a 44.2 percent addition to labor force growth to give 144 percent; 144 percent of 2.5 percent is 3.6 percent) and significantly raise energy demands above our forecast levels. On the other hand, if the long-term flow of illegal aliens does not exceed 150,000 to 250,000 a year, it will add little to economic growth or energy demand in this century.

REFERENCES

1. See U.S. Bureau of the Census, Social and Economic Statistics Administration, *Population Estimates and Projections,* Series P-25, No. 541, U.S. Government Printing Office, Washington, D.C., February 1975.

2. U.S. Bureau of the Census, *Current Population Reports,* Series P-20, No. 307, "Population Profile of the United States: 1976," U.S. Government Printing Office, Washington, D.C., 1977, p. 6.

3. Robert Warren and Jennifer Peck, "Emigration from the United States: 1960 to 1970," paper given at the meeting of the Population Association of America, Seattle, Washington, April 17-19, 1975.

4. The Attorney General, *Illegal Immigration: President's Program,* Washington, D.C., February 1978.

5. Department of Justice, Immigration and Naturalization Service, *Estimated Total Number of Illegal Aliens and Employed Illegal Aliens by I&NS District,* March 17, 1976 (mimeographed release, Immigration and Naturalization Service, Washington, D.C.).

6. David S. North, New Trans-Century Foundation, Washington, D.C.; Lesko Associates, Washington, D.C.

7. Bureau of Economic Analysis, Department of Commerce, *Population, Personal Income and Earnings By State, Projections to 2000,* Washington, D.C., October 1977, pp. 6-9.

8. Department of Justice, Immigration and Naturalization Service, *Illegal Alien Study Design,* Volume 1—Final Report, prepared by David S. North, Linton & Co., Washington, D.C., May 1975, p. 94.

9. See John E. Karkashian, *The Illegal Alien,* Department of State Senior Seminar in Foreign Policy, Washington, D.C., 1976, pp. 23-24. A collection of estimates is given in John S. Evans and Dilmus D. James, "Conditions of Employment and Income Distribution in Mexico as Incentives for Mexican Immigrations to the United States," paper presented to the North American Economics Association-Southern Economics Association, Atlanta, Georgia, November 18, 1976.

10. David S. North and Marion F. Houstoun, personal communication.

11. Series II and II-X are contained in U.S. Bureau of the Census, *Population Estimates and Projections,* Series P-25, No. 601, U.S. Government Printing Office, Washington, D.C., October 1975.

12. D. S. North and M. F. Houstoun, *The Characteristics and Role of Illegal Aliens in the U.S. Labor Market: An Exploratory Study,* Linton & Co., Washington, D.C., March 1976, pp. 5-10.

13. E. L. Allen *et al., U.S. Energy and Economic Growth, 1975-2010,* Publication ORAU/IEA-76-7, Institute for Energy Analysis, Oak Ridge Associated Universities, Oak Ridge, Tennessee, 1976.

14. One estimate is that perhaps 65,000 illegal alien children were attending New York City schools in 1973. See L. A. Westoff, "Should We Pull Up the Gangplank?" *New York Times Magazine,* September 16, 1973, p. 15.

APPENDIX B

THE FUTURE OF COAL

Introduction

The long-term competitive position of the relatively abundant coal reserves has apparently improved significantly as a result of dwindling domestic supplies of oil and natural gas. Coal is now the cheapest of the fossil fuels on a Btu basis, but the costs of burning it are rising, primarily as a consequence of federally mandated pollution controls and other regulatory measures. Whereas in 1976 coal seemed assured of a larger and rapidly growing share of the public utility and industrial fuels markets, replacing oil and natural gas, this outcome now is much less certain.

In President Carter's National Energy Plan (NEP), there is the statement (1):

Industry and utilities consumed 4.8 million barrels of oil per day and 5.9 million barrels of oil equivalent in the form of natural gas in 1976. Oil and natural gas are scarce, and generally they are needed more by other sectors of the economy. Industry and utilities can convert to other energy sources more readily than can other users; therefore a large scale conversion by industry and utilities from oil and gas to more abundant resources is needed.

The future quantities of coal consumed will depend upon (i) the interpretation of federal legislation already enacted or yet to pass the Congress and (ii) the success of research, development, and demonstration (RD&D) programs designed to make the burning of coal more acceptable environmentally.

Institute for Energy Analysis projections of base-line domestic coal demand by the year 2000 range from 31.5 to 38.8 quads in the two scenarios given in this study, compared to 13.7 quads in 1976. This is a compound annual growth of from 3.0 to 3.9 percent. It is anticipated that the fastest growth will come in the western Great Plains states and will increase the proportion of low-sulfur coal mined. However, western reserves are extensive and the rates of growth projected are well within the bounds of present manpower and equipment availabilities. The industry can be characterized as one of relatively constant cost, with ease of entry, modest capital requirements, long-term contract marketing arrangements for most of its products, and abundant reserves. Hence, the increase in output is not supply-limited in raw materials, manpower, or capital. Cost increases in mining coal are primarily a consequence of four developments: (i) higher health and safety standards for coal miners, (ii) mandated restoration of strip-mined areas, (iii) environmental and legal complications on the federal leasing of coal lands, and (iv) highly inflationary wage settlements granted to the United Mine Workers of America. Each of these developments is discussed briefly below.

Mine Health and Safety Costs

Much eastern coal must be recovered from deep mines, which historically has meant a high rate of attendant loss of life or incapacitation caused by accidents and pulmonary disease. A consequence of the Coal Mine Health and Safety Act of 1969 has been the issuance of new health and safety standards by the Department of the Interior for both underground and surface mines. One major problem is coal dust, which is linked to the incidence of black lung; benefit payments plus administrative expense under programs operated by the Social Security Administration and the Department of Labor now amount to about $1 billion a year (2). In addition, costs of coal mine accidents amounted to $57 million in 1974 (3). One federal government report estimates that, if the fatality and disability injury rate does not improve greatly from the 1975 rate, from 3400 to 4700 miners could be killed and 253,000 to 351,000 disabled in the 25-year period from 1975 to 2000 [(4), p. 4137]. The difference in these predictions represents two different estimates of coal production to 2000.

Since the passage of the Coal Mine Health and Safety Act of 1969, there has been an 8-year decline in productivity which cumulatively totaled more than 50 percent. Underground mines registered an average output per man per day of only 8.5 tons in 1976 (5). Not all of the decline in productivity may be attributable to the 1969 act, although the statistics on productivity record a decline thereafter. Declines in productivity have been reflected in higher prices since 1970. If the productivity trend continues to be negative, then the long-term competitive position of coal will weaken. A complicating factor has been the laws passed during the Carter Administration designed to slow the expansion of highly efficient western coal mines and to stimulate output in the Appalachian and midwestern states where productivity is comparatively low. This development is discussed below in the paragraphs on the best available control technology. Table B-1 compares the performance of the high-production states to Pennsylvania, even though about half of Pennsylvania's coal is strip-mined.

Strip-Mining Costs

The passage of a strip-mining law in 1977 requires, among some 25 provisions, that mining companies restore stripped land to a condition capable of supporting whatever function for which it was used prior to coal mining (7). The law sets up a $4.1 billion fund to pay for the restoration of strip-mined land already abandoned. It also includes restrictions on mining coal under agricultural land and gives farmers and ranchers veto power over mining on their land, even though the mineral rights belong to others including the federal government. Implementing regulations were announced late in 1977 and have been criticized as being far more severe than the law requires. It undoubtedly will be several years before the financial and production implications of tighter control over strip-mining will be measurable, but there is no question that mining costs will increase as a consequence.

Table B-1.
Coal mining average output (in tons per man per day), 1975 (6)

State	Output
Arizona	69.66
Montana	127.25
North Dakota	86.86
Texas	76.49
Pennsylvania	11.46

Federal Leasing of Coal Lands

The policies and actions of the federal government on the leasing of government-owned coal lands are important in the development of the industry because the western lands owned by the federal government contain an estimated 40 percent of the nation's coal reserves. Most of this coal is located on federal lands, but only a few percent of U.S. coal output in 1975 came from this source. Concern over inactive leases led to the imposition of a 2-year leasing moratorium (May 1971 to February 1973). Only a few short-term leases have been granted since then.

Formerly, it was possible for a private firm or an individual to obtain a prospecting permit to explore for commercial deposits of coal on specific locations (up to 5120 acres) of federally owned lands. The Mineral Leasing Act of 1920 provides for the issuance of such permits, but on January 26, 1976, the Secretary of the Interior decided on a leasing policy under which no further prospecting permits would be issued. Moreover, the 1976 amendments to the Mineral Leasing Act impose an additional role for government regulation, including the requirement that leases be developed within a decade (8).

The requirement for an environmental impact statement as part of the development plan submitted to the Department of the Interior's Bureau of Land Management has been an additional source of delay. Environmental lawsuits are delaying new leases as well. A spokesman for the General Accounting Office believes that the mass of restrictions will hold 1985 coal output below 1×10^9 tons, compared to the federal goal of 1.265×10^9 tons. The NEP projects that coal use by utilities in 1985 will equal 16.6 quads, compared to 9.8 quads in 1976. It is most unlikely that utilities will consume this much coal in 1985 [(4), p. 2].

United Mine Workers Wage Settlement

With the growth of western coal mining, the share of the national production accounted for by miners affiliated with the United Mine Workers (UMW) has been dropping. Currently, the UMW accounts for only slightly over half of the total production. The 1978 wage settlement was preceded by a very large wage increase in 1974. In order that wage increases be noninflationary, they are supposed to be tied to productivity increases. But, since productivity trends have been negative in coal mining, wage increases are automatically inflationary. Initial estimates are that the 1978 wage settlement will increase the miner's pay by about $4.50 an hour over the 3-year period of the contract (9). However, the impact on coal prices will depend very importantly on what happens to productivity. Unless productivity begins to improve, the incentives to shift to coal in President Carter's 1977 energy program may well be eroded.

The BACT Requirement

Apart from the four major factors discussed above which affect coal mining costs, there are additional regulatory actions which increase the cost of coal use. The best available control technology (BACT) requirement is the most important of these.

Regulations controlling the emission of sulfur dioxide (SO_2) seriously inhibit additional coal use. Environmentalists, regional political groups, state governments, and the U.S. Environmental Protection Agency (EPA) have combined to impose more stringent requirements. In 1977, the option of using low-sulfur western coal to meet environmental standards was circumscribed by the amendments to the Clean Air Act.

A major stimulus to the mining and use of low-sulfur coal in the West was the passage of the Clean Air Act Amendments of 1971, which required new coal-fired power plants built after that time to emit not more than 1.2 pounds of SO_2 per 10^6 Btu of heat input. The utilities turned to the use of so-called conforming coal, that is, coal that could be burned without scrubbers and still conform to the new federal standards. The new requirement boosted the competitive position of low-sulfur western coal, particularly from the northern Great Plains region of Wyoming and Montana. This coal quickly penetrated into the East North Central states to meet the new source performance standards. Utilities in Illinois, the largest coal-producing state in the Midwest, were drawing one-third of their coal supply from the West by 1975. Virtually no western coal was burned in Michigan and Ohio in 1973 because of the greater shipping distances, but by 1975 over 3×10^6 tons were burned by utilities in these two states (10).

For existing power plants, the individual states were given jurisdiction over the environmental standards. State implementation plans vary widely, and many states now have emission standards that are more stringent than the federal regulations.

The passage of the Clean Air Act Amendments of 1977 (11) introduced a new regulation of great regional importance to the coal industry, since it negated the competitive positions of western coal in midwestern markets. It requires the use of BACT, which today means stack scrubbers (a stack scrubber is designed to remove pollutants, such as SO_2 or particulate matter, from stack gas emissions before these gases are emitted into the atmosphere). There are two primary difficulties with stack scrubbers: (i) they do not work very well, and (ii) they are very expensive.

Section III of the Clean Air Act Amendments of 1977 [(11), p. 29] states that for new stationary power plants

. . . a standard of performance shall reflect the degree of emission limitation and the percentage reduction achievable through application of the best technological system of continuous emission reduction.

The standard of performance established by the EPA requires a fixed percentage reduction in emissions regardless of the sulfur content of the coal being burned, which effectively removed much of the economic incentive to transport low-Btu western coal long distances. The EPA has the authority to set the percentage reduction that will apply. In a draft document (12), it proposed 90 percent sulfur removal as the general requirement. This proposed standard may be eased before the issuance of a regulation. However, it is probable that any new utility plants constructed after early 1978 will need to incorporate flue gas desulfurization (FGD). The EPA is required to review the situation every 4 years and set new standards for BACT, and standards may become stiffer.

The impact of the BACT requirement is to increase utility investment costs by 20 percent or more. There is also an additional operating cost for absorbent materials, additional energy needed to operate the scrubbers, and the expense of disposing of the sludge. Overall, these cost increases would seem to favor additional nuclear-powered steam-generating plants in cases where coal-fired steam-generating plants had been comparable in cost.

RD&D and Future Coal Demand

The foregoing analysis leads us to the conclusion that coal demand is likely to fall far short of the tonnages projected in Carter Administration plans. Those who are optimistic about coal's future expect that research and development will mitigate the major environmental obstacles that now inhibit the mining and burning of coal. In President Carter's NEP, coal has been chosen to play a key role as the "swing fuel" for the rest of this century at least or until the so-called "inexhaustibles" (solar power and the breeder reactors primarily) can be brought into general commercial use. Coal represents 90 percent of the nation's conventional fuel reserves; its production could be expanded relatively quickly if it can be made environmentally acceptable. But greater coal utilization may require environmental compromises with present regulatory requirements even if RD&D programs are successful.

The energy demand scenarios of IEA for the year 2000 anticipate that electricity will expand its share of the total energy supply to 46 percent or more, compared to 28 percent in 1975, and that electric power utilities provide the largest coal market. A number of Department of Energy (DOE) research and development (R&D) programs are designed to make coal burning environmentally acceptable. These are described below. If additional large-scale expansion of direct coal burning proves to be unacceptable, a long-run alternative could be the production of synthetic gas from coal.

Coal Mining

Except for in situ gasification, there are no federally sponsored R&D efforts on coal mining of significance. There are small programs on technology and health and safety. Since productivity in deep (underground) mining has fallen from over 15 tons per man per day in the late 1960s to 8.5 tons in 1976, coal costs have risen sharply. Clearly, there is a need for new technologies for deep mining as well as the related safety equipment—improved ventilation systems and techniques for coping with various mine gases and with coal and rock dust.

Direct Combustion of Coal

Technologies in development designed to meet the problem of direct combustion applications include improved FGD and fluidized-bed combustion (FBC). The problems of the disposal of sludge from stack scrubbers and waste from FBC systems are serious and have not yet been solved. Other technologies under study are front-end coal cleaning by grinding and washing and the solvent-refined coal processes, which use chemical means to remove most of the sulfur content. The White House has stated, with respect to coal research (13), that

the highest immediate priority is the development of more effective, economical methods to meet air pollution control standards.

Flue Gas Desulfurization

The leading short-run technology is stack gas cleaning. In early 1978, there were about 30 such units in operation, an equal number under construction, and 65 in the planning stage or beyond (14). This represents significant market penetration. Operational problems have been mainly mechanical, associated with corrosion and erosion. First-generation scrubbers have worked best on low-sulfur coal (these are all lime-limestone systems). The second-generation scrubbers, some of which are in the design stage, are expected to be much improved in efficiency and reliability. Although the improved operational aspects are important, the more elaborate equipment incorporated in second-generation models means that capital costs will probably increase.

The Fluidized Bed

When fluidized-bed technologies become commercially available (these are expected before or by 1990), they could help to make coal the key transitional fuel. At present, there is no operating commercial utility plant that utilizes the fluidized bed. The three major future uses of coal, all of which would be appropriate markets for the fluidized bed, in order of probable quantities of coal consumed, are (i) electric utility boilers, (ii) industrial steam boilers, and (iii) commercial establishments (schools, hospitals, office complexes) above a minimum size.

In FBC, the fuel that is burned has earlier been mixed with some inert material and expanded (fluidized) into a relatively thick layer by the passage of air through it. At a gas velocity of three to five times the fluidizing velocity, the bed behaves like a violently boiling liquid. The bubbling action provides a high degree of particle mixing and good circulation with the exposure of a large surface area of particles.

The FBC system uses coal particles and limestone sorbent. The limestone reacts with the SO_2 produced during the combustion and is further oxidized to produce calcium sulfate which is removed with the ashes. This reaction eliminates the need to remove the oxides of sulfur (SO_x) from the stack gases.

Two varieties of fluidized-bed combustors are distinguished on the basis of whether they are "atmospheric," air being supplied to the bed at near atmospheric pressure, or "pressurized," air supplied at 4 to 10 atmospheres. As with stack scrubbers, there are problems of corrosion, erosion, and waste disposal with fluidized beds. The strategy of FBC systems development is to make coal acceptable or a substitute for oil and gas in utilities and major electricity-generating plants (15). At current rates of development, it will probably be 1985 or later before the atmospheric system is tested in a major utility mode.

The justification for a pressurized fluidized-bed system operating at 10 atmospheres rests on burning high-sulfur coal—at much increased conversion efficiencies, to be achieved with advanced power systems. Combined cycle efficiencies await the development of a gas turbine for pressurizing and for power generation. There is one report of the pressurized fluidized-bed system achieving very high SO_x removal efficiencies, on the order of 99 percent (16). The development program is in an early stage.

It was estimated in 1976 that, by the year 2000, direct combustion systems could replace between 6 and 8 quads of oil and gas [(15), p. 22]. Perhaps 2 quads of this total would be in electric utilities. This would mean an additional 300×10^6 to 400×10^6 tons of coal consumed, depending upon the proportion of low-Btu western coal in the mix. If roughly one-half of the success of advanced direct combustion systems was reflected in increased coal output (200×10^6 tons), this would be an equivalent oil saving of 2.3×10^6 barrels a day [(4), p. 9.16].

Market Penetration

One major reason for believing that coal could take over a large share of the industrial energy market, as well as the utility market, is the support of the Carter Administration for the expansion of coal use as reflected in the NEP. Under this plan, the capital cost of steam-generating facilities fired by coal are assumed to be reduced very significantly as a result of the proposed tax and rebate provisions. These provisions may be eliminated or modified by the Congress when an NEP finally emerges.

Considering the electric utility market, it is not clear that the success of FGD and FBC systems will result in any significant displacement of other fossil fuels or of nuclear-powered electricity-generating plants by 2000. Acting Energy Research and Development Administration administrator Robert Fri testified in September 1977 that, even adding the costs of waste disposal and decontamination and the decommissioning of nuclear power reactors, nuclear power would still have a cost advantage over coal-fired units (17). Projections of IEA to 2000 of electric power generation by natural gas and coal are quite small—3.1 quads in the 101- and 126-quad scenarios—and are confined to nonbase-load applications except in the case of New England.

Virtually all petroleum allocated in the IEA scenarios to industry in 2000 would be needed for feedstocks. However, after allowing for nonconvertible industrial uses of natural gas, there is an estimated 6.4 quads of convertible uses. Only about 1 quad of natural gas consumption by industry represents uses where gas would be essential.

More Advanced Technologies

In electric utility power generation, there are a number of advanced power systems still in the R&D stage which may be used if the technology of direct coal combustion proves to be unacceptable from an environmental or cost standpoint. The molten carbonate fuel cell with gasifier and the open-cycle gas turbine/integrated gasifier combined cycle are two of the more promising new coal-based electric-power generation technologies

Much of the available information on these two systems is based upon analytical studies rather than on data derived from pilot plant or industrial use (18). Production of electric power from coal-derived synthetic fuels with the open-cycle gas turbine appears promising. However, electric utilities have had problems with gas turbines used for peaking power because these units have been marketed after only a few hundred hours of prototype testing. Hence, any market penetration estimate and the associated capital costs are tentative.

Molten Carbonate Fuel Cell/Steam Power Plant

This system is believed to possess significant environmental advantages over fluidized-bed systems, with respect to SO_x, oxides of nitrogen, and particulates. High thermal efficiencies may possibly reduce the coal consumed by 25 percent compared to FGD plants. There are major uncertainties in fuel cell technology and in the complementary coal gasification system. Commercialization is not likely until the 1990s. The combined cycle systems may be available sooner or later, depending on whether emphasis is placed on higher efficiency or on more conventional technology.

Combined Cycle Turbine/Steam System

The combined cycle gas turbine/steam turbine is a promising, relatively clean, coal-based electric power-generating technology. It utilizes low-Btu gas produced at the power plant site. This gas is used to fuel a gas turbine generator system, exhausting to a heat recovery steam generator. This combination produces higher thermal efficiencies without environmental degradation. The thermal efficiencies of conventional steam power plants are limited to 38 to 40 percent. A pilot plant (Powerson) is at Pekin, Illinois (70 percent of costs have been paid by the federal government and 30 percent by private utilities) (19).

Open Cycle Magnetohydrodynamics

The open cycle magnetohydrodynamics system (MHD) in principle may be technologically attractive, since it could save about one-third on the coal requirement of a conventional plant equipped with FGD. Some aspects of the engineering feasibility of MHD have been demonstrated in small-scale tests with natural gas, but commercial operation is a long way off. The U.S.S.R. has been experimenting with this system. The environmental advantages of MHD over competitive advanced systems have yet to be established.

An MHD generator replaces both the steam turbine and the generator, and uses the combustion gases directly rather than operating through a water-heating, steam-producing unit which drives the turbine. Under the MHD principle, when gases are heated to high enough temperatures, some of the electrons are discharged from the gas molecules and the gas becomes "ionized," that is, it conducts electricity. The ionized gas can then be passed through a channel positioned in a magnetic field to induce a current flow (20).

With all coal-burning systems, there is some environmental degradation. Moreover, although carbon dioxide (CO_2) emissions may not be a direct hazard to human health, their accumulation in the atmosphere has potentially serious climatic effects. Some scientists have warned that, after CO_2 accumulates in the atmosphere to an as yet unspecified point, the warming trend will trigger global changes in climate and weather. More study is needed (21).

The environmental benefits of delaying conversions to coal may not, in themselves, be of sufficient magnitude to constitute a serious challenge to the Carter Administration's timing on the shift to coal. However, the problems of burning coal in larger quantities raise some serious questions, and both the Carter Administration and environmentalists will almost certainly be increasingly concerned with these.

REFERENCES

1. Executive Office of the President, *The National Energy Plan,* U.S. Government Printing Office, Washington, D.C., April 1977, p. 63.

2. Data received from the office of Helen Harvey, Social Security Administration, Baltimore, Maryland, November 1, 1977.

3. Bureau of Mines, news release, *Financial Costs of Coal Mine Accidents Analyzed,* Washington, D.C., April 11, 1977.

4. Comptroller General of the United States, *U.S. Coal Development—Promises, Uncertainties,* Washington, D.C., September 1977.

5. Bureau of Mines, "Production of Coal—Bituminous and Lignite—1976," *Mineral Industry Surveys,* No. 3076, Washington, D.C., 1977.

6. Bureau of Mines, "Coal—Bituminous and Lignite in 1975," *Mineral Industry Surveys,* Washington, D.C., February 10, 1977, p. 13.

7. U.S. House of Representatives, Surface Mining Control and Reclamation Act of 1977, *Conference Report,* 95th Congress, 1st Session, 1977.

8. U.S. Congress, Public Law 94-377, *An Act to Amend the Mineral Leasing Act of 1920, and for Other Purposes,* 94th Congress, 2nd Session, 1976.

9. *National Journal,* March 4, 1978, p. 355.

10. J. G. Asbury *et al., Survey of Electric Utility Demand for Western Coal,* Argonne National Laboratory, Argonne, Illinois, January 1977.

11. Committee on Environment and Public Works, U.S. Senate, *The Clean Air Act as Amended August 1977,* 95th Congress, 1st Session, 1977.

12. Environmental Protection Agency, *Standards of Performance for New Stationary Sources,* Washington, D.C., November 29, 1977.

13. Executive Office of the President, Energy Policy and Planning, *The National Energy Plan,* U.S. Government Printing Office, Washington, D.C., 1977, p. 68.

14. Bimonthly reports on FGD deployment are compiled by PEDCO-Environmental Specialists, Inc. (Cincinnati, Ohio) under EPA contract; see also Federal Power Commission, *The Status of Flue Gas Desulfurization Applications in the United States,* Washington, D.C., July 1977.

15. Energy Research and Development Administration, *A National Plan for Energy Research, Development and Demonstration: Creating Energy Choices for the Future,* 1976, vol. 2, U.S. Government Printing Office, Washington, D.C., p. 23.

16. Energy Resources Company, Inc., *Environmental Research Needs for Coal Conversion and Combustion Technologies,* Cambridge, Massachusetts, December 1976, p. 31. Fluidized-bed experts at the Department of Energy believe that this performance is unusually high.

17. Reported in *Energy Users Report,* September 1977, p. 18.

18. Assistant Administrator for Fossil Energy, *Fossil Energy Research Program of the Energy Research and Development Administration,* "Advanced Power Systems," F-1, 1978, Energy Research and Development Administration, Washington, D.C., 1977, pp. 139-147.

19. Electric Power Research Institute, *EPRI Journal,* vol. 2, No. 9, pp. 33-35. November 1977.

20. *Information from ERDA,* "Fact Sheet, Magnetohydrodynamics," vol. 3, No. 24, pp. 5-6, Energy Research and Development Administration, Washington, D.C., June 17, 1977.

21. *Present and Future Production of CO_2 from Fossil Fuels—A Global Appraisal,* Publication ORAU/IEA(0)-77-15, Institute for Energy Analysis, Oak Ridge Associated Universities, Oak Ridge, Tennessee, June 1977.

APPENDIX C

ENERGY PRICES
AND THE GROSS
NATIONAL PRODUCT

In the earlier sections of this study, the future gross national product (GNP) and energy demand have been estimated from an examination of trends in the labor force, productivity, and sectoral energy efficiencies. In these estimates we have not introduced energy prices explicitly; instead, we have given independent estimates of energy prices, based generally on extrapolation and judgment. However, implicit in our projections are price elasticities, and it is necessary to determine whether these elasticities are plausible. An analysis of future energy prices and elasticities follows in this section. We find that our elasticities are well within the range of elasticities obtained in other studies.

Energy Prices

Estimates of future energy prices over the next 25 years obviously involve great uncertainties. In the short term, unsettling events of limited duration, for example, an oil embargo, could bring sharp but temporary price increases. In the longer term, we expect average energy prices to increase more rapidly than prices in general in the economy since world energy demands will keep growing in response to population and GNP growth and more costly energy resources will be tapped and transported at higher costs to satisfy these demands.

Coal

Coal reserves in the United States are very large. Within the limits of estimated demand, coal does not appear to present a production problem over
the next 25 years. However, the lead time needed to develop large underground mines is about 4 to 5 years. Surface mines, the predominant mining
practice in the western United States, can be developed more quickly, but the
process is still time-consuming. Hence, coal supply is not particularly elastic
in the short run. In the long run, since production is not concentrated in a
few firms and the entry of new firms is not impeded by institutional constraints (patents and the like), economies of scale, or high capital costs,
equilibrium coal prices are expected to approximate the costs of production,
defined to include an economy-wide average rate of profit.

Coal prices to 1985 were estimated by the Federal Energy Administration
for individual geographic regions (1). For representative regions, these projected price increases between 1975 and 1985 are calculated to be about 22
percent, measured in constant 1975 dollars. We have used the same annual
percentage increase (2 percent) for the time beyond 1985.

Zimmerman (2) has estimated that a price increase of no more than 22 percent imposed between now and 1985 would extend the remaining life of coal
resources in the Appalachian and Illinois basins by a factor of 4. At current
rates of production, the life of the Illinois basin would be extended to 90
years and that of the Appalachian resources to 38 years if prices increase by
22 percent (projected 1985 average prices).

Given the slow growth in additional coal-fired power-generating facilities,
would the real costs of coal expansion push coal prices up faster than a net 2
percent per year? The greatest expansion of the coal industry, experts believe,
will come about in the western United States. In an Argonne National Laboratory study (3) of expansion capability for this area in the period from
1974 through 1982, the conclusion was reached that price increases will be
moderate if annual growth does not exceed 25 percent per year. This rate of
expansion is far above that required to satisfy the coal needs of our scenarios,
hence we would expect a lower rate of price increase.

Oil and Natural Gas

Domestic oil prices, on the other hand, are expected to increase substantially
between now and 1985, and to equal world oil prices by or before 1985.
Domestic price controls may be extended beyond May 1979, but a net increase in the world oil price of only 2.1 percent annually would bring it to

$16 per barrel by 1985. This is the estimated 1985 average price we have used in this analysis. Oil prices are expected to increase more rapidly thereafter as domestic production peaks (about 1985), even with the additional production from Alaskan and offshore sources. Extraction of oil in the longer term is expected to include increasingly difficult and therefore higher-cost environments, such as the Beaufort Sea.

As a bargaining tactic, the Organization of Petroleum Exporting Countries (OPEC) has insisted that future price adjustments should at least offset inflation in Western nations. During the period from 1973 through 1977, inflation in other industrialized nations ran higher than in the United States; however, in 1978 U.S. consumer prices were growing by about 8 percent annually, and so in the United States the rate of inflation exceeded that in any other Western country. Also, the most visible price increases on U.S. products are those for high-technology military materials imported into the Middle East, for which it is not possible to construct meaningful price indices. This makes price increases for oil a matter of negotiation rather than of exact determination. Under these circumstances, OPEC price demands could average 10 percent per year. Under the twin key factors expected to govern future oil prices after 1985—a drop in U.S. oil output and strong OPEC bargaining—domestic oil prices will be determined by world oil prices. In our scenario we project for the period from 1985 through 2000 a 3 percent annual increase above the rate of inflation.

Shale oil may be in commercial use during the period from 1985 through 1990. The prices of synthetics from coal are expected to be higher than that of shale oil and are not incorporated in our supply scenarios until after 2000.

Natural gas prices are assumed to increase by 1985 to $2.76 per 10^6 Btu, the equivalent Btu price for oil at $16 per barrel. The rapid price adjustment for natural gas, in the face of limited deregulation, began in 1975 when wellhead prices increased 43 percent. After 1985, natural gas prices are assumed to follow oil prices on a Btu basis.

Electricity

The future cost of electricity in the United States will depend upon (i) the mixture of electricity-generating plants in service (coal-fired or nuclear), (ii) the economic factors governing discount and inflation rates and fuel costs, (iii) the demand for electricity as a substitute for processes that now use oil and gas, and (iv) the specific regional characteristics related to energy demands and fuel supplies. The future demand for electricity is projected to grow to the year 2000 at an average annual rate of 3.5 percent for the 101-quad case and 4.8 percent for the 126-quad case.

Our estimated prices of different energy modalities are summarized in Table 6-1.

REFERENCES

1. Federal Energy Administration, *Interagency Task Force on Coal,* Project Independence Blueprint Final Task Force Report, Washington, D.C., 1974.

2. Martin B. Zimmerman, "Long Run Mineral Supply: The Case of Coal in the United States," review of Ph.D. thesis at Massachusetts Institute of Technology by Richard L. Gordon, Electric Power Research Institute, Research Project 335-1, vol. 1, February 1976.

3. J. G. Asbury and K. W. Costello, "Price and Availability of Western Coal in Midwestern Electric Utility Market, 1974-1982," Argonne National Laboratory, Argonne, Illinois, October 1974 (distributed by the National Technical Information Service, Springfield, Virginia).

INDEX